Food Legislation of the UK

ONE WEEK LOAN

Food Legislation of the UK

A concise guide

Fourth edition

D. J. Jukes BSc, MSc, PhD, MIFST

Lecturer in Food Technology
Department of Food Science and Technology
University of Reading

Butterworth-Heinemann
Linacre House, Jordan Hill, Oxford OX2 8DP
A division of Reed Educational and Professional Publishing Ltd

 A member of the Reed Elsevier plc group

OXFORD JOHANNESBURG BOSTON
MELBOURNE NEW DELHI SINGAPORE

First published 1984
Second edition 1987
Third edition 1993
Revised reprint 1994
Fourth edition 1997

British Library Cataloguing in Publication Data
A catalogue record for this book is available from the British Library

ISBN 0 7506 3385 9

Typeset by Avocet Typeset, Brill, Aylesbury, Bucks
Printed and bound in Great Britain by Biddles Ltd, Guildford and King's Lynn

Contents

For my wife Helen and our daughters Camilla, Polly, Verity, Kezia and Leah.

Important

Readers should note that this Guide is not a legal document and, whilst every care has been taken in its preparation, no responsibility can be accepted for problems arising from its use. Only the Courts can provide an exact interpretation of the legislation.

Acknowledgements

The inspiration for this Guide came from the original work of David Pearson whose *Concise Guide to UK Food Legislation* was published in May 1976.

I would welcome comments from readers on the content, presentation and relevance of any of the topics covered (or left out).

David Jukes
Lecturer in Food Technology
Department of Food Science and Technology
Food Studies Building
University of Reading
Whiteknights
Reading RG6 6AP

1 Introduction

This guide has been designed to cover the legislative controls under which food technologists have to work in the UK. The much smaller original document was prepared to provide students of the then National College of Food Technology at Reading University with sufficient detail of the legal standards they would need when working as technologists in industry. Now into its fourth edition, continued substantial revision has been required to incorporate new material and to amend many older controls. Despite attempts at deregulation, the quantity of material covered has grown considerably. Its continued publication illustrates that a wide range of people require guidance, in a summary form, of information on the laws and regulations applied to food and food processing.

The selection of regulations for inclusion continues to be difficult and readers should note that this guide does not provide a comprehensive list of all the regulations affecting food and food production. In particular, a distinction has been drawn between the requirements for food products, which are mostly covered, and the requirements for agricultural produce, which are not. Thus, for example, grading standards for vegetables and eggs are not included. A further distinction is made between food regulations of interest to the industry as a whole, and those complex regulations affecting one particular section. As an example of this, the EEC Wine Regulations have not been included. Regulations affecting all industry and not specifically food production are also not included. In a few cases, the complexity of the controls has been such that producing a summary version has been considered to be inappropriate – in these cases, some background information is given, but readers are referred to the Regulations (pesticide residues and labelling of chocolate products are two examples of these).

It should be noted that this guide is based on the regulations for England and Wales. Whilst most of the regulations for Scotland apply the same standards, certain details may be different. Appendix 1 should be consulted for further details. A similar situation affects the controls for Northern Ireland.

As with any attempt to simplify legal matters, it is likely that in the process certain requirements or exemptions will have been dropped. It is

1

1

important that readers appreciate that this guide is not a legal document, and that for full details of the legal requirements the acts and regulations must be consulted.

Legislation is constantly changing. This guide gives the legal requirements at the time of writing. Any changes that occur prior to printing but after the preparation of the main text will be included in Appendix 6. It is therefore suggested that readers check Appendix 6 prior to using the guide. It is also the author's intention to provide amendment information on the internet. Readers with access to the internet are welcome to check the page at the following address: http://www.fst.rdg.ac.uk/people/ajukesdj/book.htm.

2   UK food law

In the UK, as elsewhere, legislation controlling food has two principal aims – the protection of the health of the consumer and the prevention of fraud. In addition there is legislation which has an economic motive (tariffs, taxes, quotas, etc.), which provides economic benefits to an industry or country but which does not usually involve the technologist. The two principal aims are achieved in the UK by a combination of primary legislation (the acts) and more detailed secondary legislative measures (the regulations or orders). The acts usually contain general prohibitions, which, when interpreted and enforced by the courts, provide general consumer protection. The scientific and technical requirements of food production require more detailed controls. The acts therefore also contain the authorisation for Ministers, subject to specified parliamentary approval, to issue detailed regulations. It is these which form the main bulk of this book.

While this book is meant as a guide to the technical requirements of legislation, it is necessary to put it into the context of the legislative structure in the UK. It is therefore important to discuss briefly some of the more important aspects of the application of food law in the UK.

Primary legislation

In England, Wales and Scotland, the primary legislative powers are now contained in the Food Safety Act 1990. This act updated the primary legislation for England, Wales and Scotland. For Northern Ireland, very similar controls are contained in the Food Safety (Northern Ireland) Order 1991. Only very brief details of the act are given in this book and readers are strongly urged to obtain an actual copy of the act.

The act states that food should not be 'rendered injurious to health' and should comply with certain specified 'food safety requirements'. General consumer protection is contained in a section which requires food to be of the 'nature or substance or quality demanded by the purchaser' and another which makes it an offence for food to be falsely described or to have a misleading label. There are numerous provisions providing for more detailed controls to be applied by regulations, and

3

2

these can cover a wide range of topics: composition, labelling, processing, packaging, hygiene, microbiological standards, novel foods, etc.

Enforcement powers are detailed and enable enforcement officers to enter premises to inspect the business to ensure compliance with the various controls. If they suspect a failure to comply, food can be seized, instructions can be issued to improve the premises ('improvement notices') and, in extreme cases, the business can be prevented from operating ('emergency prohibition notice') pending a court hearing. Failure to comply with the provisions of the act is a criminal offence and courts have significant penalty powers. An opportunity is provided for a food business to defend itself by using certain statutory defences provided by the act. In particular, provision is made for a defence if the person charged can show that they 'took all reasonable precautions and exercised all due diligence'. However, this requires the presentation of sufficient information to convince the court that the requirements have been met. It is not an easy defence to operate. For example, what is 'reasonable ' will vary from one business to another depending on the size of the operation and the potential hazards of the products.

A number of other acts apply to food products and therefore require consideration. In particular the Weights and Measures Act 1985 (a consolidation of previous acts) has many provisions which apply to food. Other acts have to be considered and are listed in Section 3.3. Special mention should be made of the Food and Environment Protection Act 1985 which contains provisions enabling the prohibition of food sales where health hazards may exist and for the control of pesticides (and their residues).

Regulations

The vast majority of the detailed technical requirements for food products are contained in the regulations issued by Ministers and laid before Parliament. In most cases the regulations are passed unless Parliament specifically votes against them. In other cases draft regulations may have to be presented to Parliament a number of days before the actual regulations are published and, in a few cases, Parliament has to positively approve the regulations. Prior to the Food Safety Act 1990, since there were three separate acts covering the UK, there had to be three separate sets of regulations issued on any particular subject. For most products this does not present any difficulty since the technical requirements are identical and only certain administrative provisions may differ. However, for historical reasons the legislation on milk for Scotland has evolved slightly differently and certain technical standards are different from those specified for England and Wales. Provisions for food hygiene that include certain registration requirements also differ. The situation has now been improved to the extent that England, Wales and Scotland can have combined regulations since the Food Safety Act 1990 applies to all three countries. Details are given in Appendix 1 of those regulations which are similar and those which contain some differences in detail.

Depending upon the requirements of the primary legislation, some of the secondary powers are issued as 'orders' rather than 'regulations'. For England, Wales and Scotland these are all published as Statutory Instruments (SIs) with a reference number indicating the year of publication and the number assigned to the particular SI. The equivalent regulations for Northern Ireland are generally issued in a separate series known as the Statutory Rules of Northern Ireland. Due to the difficult legislative position in Northern Ireland, certain measures (including primary legislation) are being published as Statutory Instruments. It is therefore important to check which series is involved when quoting a reference number.

Development of regulations

Before issuing regulations, Ministers are usually required to consult with those organisations which may be affected by them. Thus, when a Minister proposes to make a regulation, a press notice is issued and the proposed regulation is sent for comment to manufacturers, retailers, consumer groups and a number of professional bodies representing technologists, enforcement officers, councils, etc. Only when this consultation is complete will the Minister decide on the final form of the regulations. For matters which require more extensive or formal scientific consideration a number of independent committees have been established to advise the Ministers. The principal one for food is the Food Advisory Committee (FAC) which was constituted in 1983 by combining two previous committees (the Food Standards Committee (FSC) and the Food Additives and Contaminants Committee (FACC)). When requested to study a particular topic, the FAC invites comment from interested parties and often asks other committees to give advice on certain specialised subjects. The publication of a report by the committee is often used as the basis for future regulations. A full list of published reports is given in Appendix 2. In addition, many of the committees are producing annual reports which give a detailed account of all the activites of the committee during the preceding year

Compositional standards

One of the methods used to prevent consumer fraud is to ensure that products described by a particular name conform to certain standards.

2

When standards are set for a particular name then that name is known as a 'reserved description' and can only be used for products conforming to the set standard. When someone buys that product the legislation ensures that the consumer is buying a set standard and hence a satisfactory product. Compositional standards were originally permitted by the Food and Drugs Act of 1938, but the stimulus for such standards was provided by the need for controls during wartime rationing. Many of these were continued after the war and additional compositional standards were added. Some rather obscure products were included in this category (mustard, curry powder and tomato ketchup, for example) as well as some of the staples of the British diet (flour, bread and sausages). During the 1950s and 1960s compositional controls were considered the most useful way of providing consumer protection.

Soon after the UK joined the European Common Market in 1973 further regulations were issued covering compositional standards for a number of products (cocoa and chocolate, sugar products, fruit juices and nectars) some of which had never previously had a UK standard. However, more recently the emphasis has been to restrict the number of compositional standards to a number of basic products and enable manufacturers to produce a broad spectrum of goods. Consumer protection is maintained by more stringent labelling requirements. This trend is demonstrated by the revocation, in 1990, of the legal standards for mustard, curry powder and tomato ketchup and, in 1995, of the controls on bread, cream and cheese.

Additives

The use of most additives in food products is controlled by regulations which adopt the 'positive list' system. Under this system only those additives positively listed in the regulations which perform the function (antioxidant, colour, etc.) defined by the regulations are permitted for food use. Provision is made in the definitions to cover the case of additives having a number of different functions. When it is considered necessary to restrict the use of a particular additive, the regulations also contain provisions that limit the foods in which the additive may be used, and the levels of usage.

All additives are subjected to a number of evaluations prior to being positively listed. Within the UK, the initial assessment is on the basis of any technical need for the additive and is conducted by the FAC. If they are satisfied that the additive does serve a useful purpose, then they ask the advice of other committees, in particular the Committee on Toxicity of Chemicals in Food, Consumer Products and the Environment (COT). These committees examine all the evidence relating to the toxicological evaluation of the additives before advising on their safety for food use.

All major categories of additives are controlled by regulations – based on European Union (EU) agreements. The main categories are covered in three sets of regulations issued at the end of 1995 and covering sweeteners, colours and miscellaneous additives (including preservatives, antioxidants, emulsifiers and stabilisers). There are separate controls on flavourings, which were applied in 1992. In this case, given the great number of chemicals involved, a 'positive list' has not yet been established and the regulations contain a 'negative list' which places restrictions on a few substances known to be hazardous.

2

Contaminants

The controls on specific contaminants in the UK are of a limited nature. Government policy has been to monitor and, when necessary, agree a voluntary policy of eliminating contamination at source rather than establishing numerous different levels for different contaminants in different foods. Only where a hazard has been sufficiently great or where the contaminant is widespread have controls been introduced. Thus, maximum arsenic, lead and tin levels have been introduced by regulation and, more recently, maximum aflatoxin levels on nuts and nut products sold to the ultimate consumer have been imposed.

The control of pesticide residues had also been by a monitoring process. Detailed regulations have now been issued in line with EU requirements and using the powers in the Food and Environment Protection Act.

It is important to note that prosecutions can be brought under the Food Safety Act on the basis that a product is injurious to health and that, in establishing this fact, non-statutory limits (internationally recommended limits, for example) can be used as evidence in the court. It is then up to the court to decide whether to accept the evidence as valid.

Processing and packaging

Under this category are included certain specific regulations that are inappropriate elsewhere. Examples include irradiation, permitted since January 1991, and controls on quick-frozen foodstuffs (implementing an EC Directive). Controls on packaging materials have also been established to prevent any harmful substances becoming incorporated into food.

2   Labelling

Labelling requirements for food products can be very complex. Certain products have their own regulations which incorporate labelling provisions. Most food products though are subject to the Food Labelling Regulations 1996. Any changes to these regulations can have quite major consequences for the whole food industry. These new regulations have incorporated moves to reduce the number of detailed compositional standards. They now contain the labelling requirements for cheeses, cream and ice-cream which had previously been contained in detailed compositional controls.

Hygiene and health

The maintenance of satisfactory hygiene practices is ensured by the various food hygiene regulations. While these are the most used regulations in terms of food prosecutions, the vast majority of cases are against small retailers, restaurants and catering establishments.

The regulations have certain specific requirements but in most cases they establish the general principles and it is for the enforcement authorities and the courts to interpret the requirements. These have been altered since the third edition of this Guide, and now are based on EU-agreed measures for food hygiene. This has now introduced the concept of hazard analysis and critical control points (HACCP) which, although not obligatory, is strongly recommended in guidance documents.

Associated with hygiene are the various regulations governing public health with regard to milk handling, poultry and meat inspection, and the various import controls to prevent the spread of certain specific animal and human health hazards. The possibility of the spread of disease from country to country has always been a risk. Even within the UK powers exist to control food movements and, in particular, Northern Ireland has certain powers to restrict the import of food from the mainland of England, Wales and Scotland. These controls have been relaxed but it is wise to check the current position with the Northern Ireland authorities.

Where it has been considered important to prevent the spread of disease by incorporating a heat-treatment step in the processing of food, standards have been incorporated in regulations. Thus a number of different heat treatments are permitted for milk so as to eliminate pathogenic organisms. Other products with similar controls are dairy products, egg products and ice-cream.

Microbiological standards for food products is another area in which the UK authorities have believed that statutory limits are usually inappropriate. As with contaminants, the emphasis has been on monitoring and ensuring good hygiene practice rather than establishing limits for specific products. However, many European countries do have such statutory standards and it is possible that the UK will eventually have more. It is worth noting that both the Natural Mineral Waters Regulations and the Dairy Products (Hygiene) Regulations do incorporate certain microbiological limits.

Weights and measures

Although the subject of weights and measures legislation is vast and mostly beyond the scope of this book, there are two aspects which do regularly affect the food technologist. These are the controls on specified weights (prescribed weights) which restrict pack sizes on a number of food products to certain set weights and the Packaged Goods Regulations which apply an average weight system to most food products. Weights and measures legislation is the responsibility of the Department of Trade and Industry.

Enforcement structure

Enforcement of food legislation in the UK has, since its start in the late nineteenth century, been the responsibility of local authorities. Each local authority employs trained personnel to ensure that a wide range of acts and regulations (including food) are enforced. The two major types of trained officers involved in food work are the Environmental Health Officer (EHO) and the Trading Standards Officer (TSO). Veterinary supervision is also required in certain areas where animals enter the food system (e.g. inspection at time of slaughter).

Environmental health officers usually have responsibility for enforcing those aspects of food law which have a hygiene or health basis. Their main area of work is therefore the enforcement of the food hygiene regulations and the controls on unfit food contained in the Food Safety Act. Trading standards officers are involved in the enforcement of a wide range of controls on all types of trading. Thus for food they usually have authority relating to composition and labelling. Qualified TSOs are also the inspectors for the weights and measures legislation.

The structure of local government in the UK is very varied and thus the administration of food law varies throughout the country. For the non-metropolitan parts of England and Wales, trading standards are allocated to the county councils while the environmental health controls are allocated to the district councils (there being a number of districts in each county). In the metropolitan areas (and increasingly in new 'unitary

2

authorities'), where there are no counties, both responsibilities are covered by the metropolitan district councils (or for London, the borough councils). For Scotland, trading standards and environmental health work is now covered by the district councils. In Scotland there is an important difference in that, since April 1983, responsibility for food composition and labelling has been enforced by EHOs. For Northern Ireland, food law administration is the responsibility of a number of Group Public Health Committees. Imported food arriving in the UK is usually inspected initially by officers employed by the Port Health Authority, although inspection may be delayed until the food arrives at its destination. Assisting in the work of the enforcement officers is the public analyst who is usually appointed by the county council to provide official analysis of samples taken by the officers. Their analysis, usually chemical or physical, is then used as the prime evidence in any prosecution relating to the composition of food. Microbiological examination of food is undertaken by food examiners who are required to be qualified to certain minimum standards. The concept of microbiological 'examination' was introduced for the first time in the Food Safety Act 1990.

With the very varied structure of local government and the enforcement of legislation at a local level there has always been scope for a different degree of enforcement around the country. Thus, even if a food label has been agreed by one trading standards department, those in another part of the country could still object to it. The need for uniform interpretation and application of the law is therefore important. To help achieve this, trading standards departments and environmental health departments are all represented on a national body – the Local Authorities Co-ordinating Body on Food and Trading Standards (LACOTS). This organisation provides a valuable forum for discussion among TSOs and EHOs and hence helps to ensure that the law is applied in a uniform way. In particular they have established a system where most companies only have to deal with one local authority (the 'home authority') and any complaints are discussed with that authority. It must be stressed though that it is possible for any trading standards department to bring a prosecution against any product sold in its area and it is for the courts to decide what the law actually requires.

The need for stronger central guidance on the execution, by enforcement officers, of their responsibilities was accepted when the Food Safety Act was passed. Section 40 of the Act provides the Minister with the authority to issue codes of recommended practice. Several have now been issued and are listed in Appendix 3.

Prior to the formation of LACOTS, another body had attempted to introduce certain nationally agreed standards in the form of codes of practice. The Local Authorities Joint Advisory Committee (LAJAC) on Food Standards produced a number of codes which are listed in

Appendix 3. While LACOTS does not actually develop codes, it has accepted a number (also listed in Appendix 3) as being useful and circulated them to its members. These codes have no legal status but they do provide a guide which can be considered as evidence by a court in any prosecution.

The European dimension

Since 1973, the UK has been a member of the EU and has therefore also been applying legislation agreed among the EU member states. European legislation is aimed at creating a common market in goods so that products produced anywhere in the Union can circulate within it unrestricted by both tariff and non-tariff barriers. EU legislation applied to food usually takes two forms – the regulation and the directive. The regulation is usually applied to primary agricultural products and is applicable to all member states at the same time. The vast majority relate to the Common Agricultural Policy and are involved with standards for intervention goods, tariffs applied to goods entering the Union and various production quotas. They are mostly beyond the scope of this book. The directive has been widely-used to try to create a market free of non-tariff barriers. Thus, differences in technical standards contained in national legislation hinder the movement of goods between member states. By agreeing common standards, these barriers are removed. Directives require implementation by each member state before they become law. Thus a large number of the UK regulations listed in this book contain standards agreed at a European level. A list of the main European legislation on food products is given in Appendix 4.

With the completion of the 'internal market' programme of the EC at the end of 1992, many more regulations and directives were adopted. Most of these have been incorporated in UK legislation. Some were adopted some time after the '1992' deadline – progress on agreeing the new controls on additives was particularly difficult. Many changes are still likely before a true internal market will exist. There remains constant pressure for new and amended controls. Technologists, and others, must remain vigilant in a constantly changing control system.

3   Acts

3.1   3.1 Food Safety Act 1990

Note: This Act is generally applicable to England, Wales and Scotland. For Northern Ireland, see: Food Safety (Northern Ireland) Order (1991/762).
Details are given of those Sections which are particularly relevant to food processing.

Part I Preliminary
Section 1   Meaning of 'food' and other basic expressions
Definition of 'food':
1) In this Act 'food' includes
   a) drink;
   b) articles and substances of no nutritional value which are used for human consumption;
   c) chewing gum and other products of a like nature and use; and
   d) articles and substances used as ingredients in the preparation of food or anything falling within this subsection.
2) In this Act 'food' does not include
   a) live animals or birds, or live fish which are not used for human consumption while they are alive;
   b) fodder or feeding stuffs for animals, birds or fish;
   c) controlled drugs within the meaning of the Misuse of Drugs Act 1971; or
   d) subject to such exceptions as may be specified in an order made by Ministers: (i) medicinal products within the meaning of the Medicines Act 1968 in respect of which product licences within the meaning of that Act are for the time being in force; or (ii) other articles or substances in respect of which such licences are for the time being in force in pursuance of orders under section 104 or 105 of that Act (application of Act to other articles and substances).
Also included are definitions of 'business', 'commercial operation', 'contact material', 'food business', 'food premises', 'food source' and 'premises'.

3.1

Section 2 Extended meaning of 'sale', etc.
This section states that the supply of food, otherwise than on sale, in the course of a business shall be deemed to be a sale. Offering food as a prize or reward or if it is given away in connection with any entertainment to which the public are admitted or in an advertisement is regarded as exposing for public sale.

Section 3 Presumptions that food intended for human consumption
The sale of food commonly used for human consumption shall be presumed to be a sale for human consumption until the contrary is proved. Any food commonly used for human consumption and found on a premises used for the preparation, storage or sale of that food, and any article or substance commonly used in the manufacture of food for human consumption which is found on such premises, shall be presumed (until the contrary is proved) to be intended for food use.

Section 4 Ministers having functions under the Act

Section 5 Food authorities and authorised officers

Section 6 Enforcement of Act

Part II Main Provisions

Section 7 Rendering food injurious to health
Any person who renders food injurious to health by means of
a) adding any article or substance to the food;
b) using any article or substance as an ingredient in the preparation of the food;
c) abstracting any constituent from the food; and
d) subjecting the food to any other process or treatment, with intent that it shall be sold for human consumption,
shall be guilty of an offence. Provision is made for any cumulative effects of consumption in ordinary quantities. Injury includes any impairment, whether permanent or temporary.

Section 8 Selling food not complying with food safety requirements
It is an offence to sell, offer to sell, deposit for the purpose of sale, etc., any food which fails to comply with food safety requirements. Food fails to comply with food safety requirements if:
a) it has been rendered injurious to health;
b) it is unfit for human consumption; or
c) it is so contaminated (whether by extraneous matter or otherwise) that it would not be reasonable to expect it to be used for human consumption in that state.
Additional provisions provide for a batch, lot or consignment to be regarded as failing (until the contrary is proved) if a part of it fails.

Section 9 Inspection and seizure of suspected food
Authorised officers are permitted to inspect at all reasonable times any food intended for human consumption which has been sold or is in the possession of a person for the purpose of sale or preparation for sale.

3.1

Additional provisions provide for powers to prevent the movement of food or to seize food if an officer considers that the food fails to comply with food safety requirements or if it appears that the food is likely to cause food poisoning or any disease communicable to human beings.

Section 10   Improvement notices

If an officer has reasonable grounds for believing that the proprietor of a food business is failing to comply with certain regulations (those which require, prohibit or regulate the use of any process or treatment in the preparation of food, or which secure the observance of hygienic conditions and practices in connection with commercial operations with respect to food or food sources) he may serve an 'improvement notice' on the proprietor. The notice will state:

a) the grounds for believing a failure to comply;
b) the matters which constitute the failure;
c) the measures necessary to comply; and
d) a time by which the measures should be taken (not less than 14 days).

Section 11   Prohibition orders

Two types of 'prohibition order' can be imposed.

1) If a court convicts a proprietor for an offence under certain regulations (identical to those listed in Section 10) and it is satisfied that there is risk of injury to health (the 'health risk condition') by
   a) the use of any process or treatment,
   b) the construction of any premises or use of equipment, or
   c) the state or condition of any premises or equipment,
   then it can make an order prohibiting, as appropriate, the use of the process, treatment, premises or equipment.

2) If a court convicts a proprietor for an offence under certain regulations (identical to those listed in Section 10) and it thinks it proper, then it can make an order prohibiting the proprietor participating in the management of any specified food business.

Section 12   Emergency prohibition notices and orders

If an officer is satisfied that the 'health risk condition' (see Section 11) is fulfilled and that there is imminent risk of injury to health, then he may serve an 'emergency prohibition notice' on the proprietor prohibiting, as appropriate, the use of a process, treatment, premises or equipment. Application must be made within 3 days to a court for the notice to be made into an 'emergency prohibition order'.

Section 13   Emergency control orders

If it appears to the Minister that the carrying out of commercial operations with respect to food, food sources or contact materials (of any class or description) involves or may involve imminent risk of injury to health, he may issue an 'emergency control order' prohibiting the carrying out of such operations with respect to food, food sources or contact materials (of that class or description).

Section 14 Selling food not of the nature or substance or quality demanded

3.1

Any person who sells to the purchaser's prejudice any food which is not of the nature or substance or quality demanded by the purchaser shall be guilty of an offence.

Section 15 Falsely describing or presenting food

Any person who gives or displays food with a label (whether or not attached to or printed on the wrapper or container) which

a) falsely describes the food; or

b) is likely to mislead as to the nature or substance or quality of the food,

shall be guilty of an offence. A similar provision applies to publishing an advertisement. Any person who sells any food the presentation of which is likely to mislead as to the nature or substance or quality of the food shall be guilty of an offence.

Sections 16–19 Regulations

These sections contain the provisions which allow Ministers to issue necessary regulations for a very wide range of purposes.

Section 16 provides the power for most common situations.

Section 17 provides additional powers for any EU obligation.

Section 18 allows for the control of novel foods or food sources and the use of genetically modified food sources.

Section 19 allows for the registration or licensing of food premises.

Sections 20–22 Defences

Section 20 provides that, where the commission of an offence by any person is due to an act or default of some other person, that other person shall be guilty of the offence.

Section 21 provides a defence for a person charged with an offence to prove that he took all reasonable precautions and exercised all due diligence to avoid the commission of the offence by himself or by a person under his control. Additional subsections of Section 21 provide for certain conditions under which a person is deemed to have satisfied this requirement.

Section 22 provides a publisher with a defence if he publishes an advertisement in the ordinary course of business and the advertisement is contrary to provisions of the Act.

Sections 23–26 Miscellaneous and supplemental

Part III   Administration and enforcement

Section 27 Appointment of public analysts

Section 28 Provision of facilities for examinations

Section 29 Procurement of samples

Section 30 Analysis, etc., of samples

Section 31 Regulation of sampling and analysis, etc.

Section 32 Powers of entry

3.1

Part IV Miscellaneous and supplemental

Section 40   Power to issue codes of practice

For the guidance of food authorities, the Minister may issue codes of recommended practice as regards the execution and enforcement of this Act and of regulations and orders made under it. Food authorities shall have regard to the provisions of the codes and shall comply with any direction which is given by the Minister and required them to take any specified step in order to comply with the code.

Section 41   Power to require returns

Section 42   Default powers

Section 43   Continuance of registration or licence on death

Section 44   Protection of officers acting in good faith

Sections 45–47   Financial provisions

Sections 48–50   Instruments and documents

Sections 51–52   Amendments of other Acts

Section 53   General interpretation

Some additional definitions are given in this section.

Section 54   Application to Crown

Sections 55–56   Water supply

Section 57   Scilly Isles and Channel Islands

Section 58   Territorial waters and the continental shelf

Section 59   Amendments, transitional provisions, savings and repeals

Section 60   Short title, commencement and extent

3.2 Weights and Measures Act 1985

Note: The majority of this Act only applies to England, Wales and Scotland. For Northern Ireland, similar controls are contained in: Weights and Measures (Northern Ireland) Order 1981 (1981/231).

The titles of the individual Parts are given below along with fuller details of those Sections having particular relevance to food.

Part I Units and standards of measurement

Section 1   Units of measurement

For measurement of mass, the pound or kilogram shall be the unit to which any measurement is made and the pound shall be exactly 0.45359237 kg.

Part II Weighing and measuring for trade

3.2

Part III Public weighing or measuring equipment

Part IV Regulation of transactions in goods

Section 22   Orders relating to transactions in particular goods

Orders may be made with respect to any specified goods to ensure that they are packed or sold in specified quantities, marked in specified ways or satisfy any other circumstance detailed in the Section.

Section 25   Offences in transactions in particular goods

Section 28   Short weight

Any person who, in selling or purporting to sell any goods by weight or other measurement or by number, delivers or causes to be delivered to the buyer

a) a lesser quantity than that purported to be sold, or

b) a lesser quantity than corresponds with the price charged,
   shall be guilty of an offence.

Sections 33–37   Defences

1) Requirements necessary for a defence of warranty are given.

2) It shall be a defence for the person charged to prove that he took all reasonable precautions and exercised all due diligence to avoid the commission of the offence.

3) If the loss occurred after packing, it is a defence if it can be shown that the loss occurred due to factors for which reasonable allowance was made.

4) For food (not prepacked) which loses weight by evaporation or drainage, if product is light, it shall be a defence to have taken due care and precaution to minimise the loss.

5) If the offence is of excess weight, it shall be a defence to prove that the excess was necessary to avoid a deficiency.

6) A person shall not be charged based on a single article unless, when available, a reasonable number of articles are also tested, and the court

   a) must consider the average weight of tested articles,

   b) if the proceedings are for a single article, shall disregard any inconsiderable deficiency or excess,

   c) have regard generally to the circumstances of the case.

Part V Packaged goods

Sections 47–48   Duties of packers or importers as to quantity and marking of containers

States the requirements of an 'average weight' system and allows for regulations to prescribe methods for the selection and testing of packages. Allows for the publishing of a code of practical guidance. States requirements for marking of packages with weight and name and

17

3.2

address or mark to enable inspector to ascertain packer or person who arranged for the packages to be made.

Sections 55–62   Co-ordination of control

These sections established a National Metrological Co-ordinating Unit with certain duties and powers. The unit has now been abolished and most of the duties and powers have now been taken over by the Secretary of State. The duties include

a) To review the operation of this part of the Act.
b) To provide information on the operation of this part of the Act.
c) To give advice to local authorities on the discharge of their duties.
d) To collaborate with similar bodies abroad on matters connected with the unit.
e) To make and maintain a record of the names and addresses of:
 i) packers and importers with the kinds of packages they make up or import and the marks they use,
 ii) makers of measuring container bottles and the marks they use.

The Secretary of State has powers to require local authorities to conduct checks on certain packages and provide information.

Part VI Administration
Section 69   Local weights and measures authorities

The local weights and measures authorities shall be the councils of the non-metropolitan counties, metropolitan districts and London boroughs.

Section 72   Appointment of inspectors

Each local weights and measures authority shall appoint a chief inspector of weights and measures and other inspectors of weights and measures.

Section 73   Certificate of qualification to act as inspector

Part VII General
Section 92   Spelling of 'gram', etc.

Permits the use of 'gramme' or '-gramme' as an alternative to 'gram' or '-gram'.

3.3 Miscellaneous acts

Prevention of Damage by Pests Act 1949

Note: Not applicable to Northern Ireland.

This Act provides for the control of pests which are infesting food premises. Infestation is defined as 'the presence of rats, mice, insects or mites in numbers or under conditions which involve an immediate or potential risk of substantial loss of or damage to food'. The following Sections have particular relevance to food processors.

Part I Rats and mice

Part II Infestation of food

Section 13   Obligation of certain undertakers to give notice of occurrence of infestation

Every person whose business consists of or includes the manufacture, storage, transport or sale of food shall notify the Minister if he is aware of any infestation present in

a) any premises, vehicle or equipment used for the above purposes, or
b) any food in his possession or other goods likely to be in contact with food.

Similar obligations apply to people handling containers intended for food use. Regulations may be issued to relax the requirements of this Section (see below).

Section 14   Powers to give directions

Directions may be given to prohibit or restrict

a) the use for food of premises, vehicles or equipment; and
b) the delivery, acceptance, retention or removal of food, which is or is likely to be infested.

Part III Supplemental

Section 22   Powers of entry

Any duly authorised person may enter any land to ascertain whether there is any failure to comply with the Act or Directions issued under the Act. The following Regulation has been issued under Section 13 of the Act: The Prevention of Damage by Pests (Infestation of Food) Regulations 1950 (1950/416). These Regulations relax the requirements of notification for infestation by insects and mites where:

a) the premises are used wholly or mainly for certain specified processes using certain specified imported foods,
b) the food is oil seeds being moved from one extracting mill to another for the purpose of oil extraction,
c) the food is fresh fruit or green vegetables, fresh home-killed meat, or fish other than cured or processed.

The Regulations should be consulted for full details.

Trade Descriptions Act 1968

Note: Applicable to all UK.

Section 1   Prohibition of false trade description

Any person who, in the course of a trade or business,

a) applies a false trade description to any goods; or
b) supplies or offers to supply any goods to which a false trade description is applied

3.3

shall, subject to the provisions of the Act, be guilty of an offence.

Section 2   Trade description

1) A trade description is an indication, direct or indirect, and by whatever means given, of any of the following matters with respect to goods or parts of goods, that is to say:

   a) quantity, size or gauge;
   b) method of manufacture, production, processing or reconditioning;
   c) composition;
   d) fitness for purpose, strength, performance, behaviour or accuracy;
   e) any physical characteristics not included in the preceding paragraphs;
   f) testing by any person and results thereof;
   g) approval by any person or conformity with a type approved by any person;
   h) place or date of manufacture, production, processing or reconditioning;
   i) person by whom manufactured, produced, processed or reconditioned;
   j) other history, including previous ownership.

The application of a legally required name under Regulations under the Food Safety Act 1990 is not a trade description

Section 3   False trade description

1) A false trade description is a trade description which is false to a material degree.

2) A trade description which, though not false, is misleading, that is to say, likely to be taken for such an indication of any of the matters specified in Section 2 of the Act would be false to a material degree, shall be deemed to be a false trade description.

3) Anything which, though not a trade description, is likely to be taken for an indication of any of those matters and, as such an indication, would be false to a material degree, shall be deemed to be a false trade description.

4) A false indication, or anything likely to be taken as an indication which would be false, that any goods comply with a standard specified or recognised by any person or implied by the approval of any person shall be deemed to be a false trade description, if there is no such person or no standard so specified, recognised or implied.

Section 4   Applying a false trade description

Section 5   Trade descriptions used in advertisements

Section 16 Prohibition of importation of goods bearing false indication of origin

Section 24 Defence of mistake, accident, etc.

International Carriage of Perishable Foodstuffs Act 1976

3.3

This Act enables the UK to accede to the Agreement on the International Carriage of Perishable Foodstuffs (the ATP). A minor amendment, not relevant to food, was made by The International Carriage of Perishable Foodstuffs Act 1976 (Amendment) Order 1983 (1983/1123). The Act enables regulations to be made governing all aspects of the carriage of perishable foodstuffs. For relevant details of the regulations, see Section 4.5.

Animal Health Act 1981

Note: Applicable to England, Wales and Scotland. For Northern Ireland, similar controls are contained in: Diseases of Animals (Northern Ireland) Order 1981 (1981/1115)

This Act consolidated several previous Acts and, in particular, The Diseases of Animals Act 1950. Section 1 provides the Minister with powers to make orders for the purpose of preventing the spreading of disease. Section 10 specifically provides for Orders for the purpose of preventing the introduction or spreading of disease into or within Great Britain through the importation of: animals and carcases; carcases of poultry and eggs; and other things (whether animate or inanimate) by or by means of which it appears that any disease might be carried or transmitted. For relevant details of the orders, see Section 4.5.

Food and Environment Protection Act 1985

Note: Applicable to all UK.

This Act covers three main areas: powers to make emergency orders when escaped substances may have contaminated food; matters related to dumping at sea; and the control of pesticides previously subject to voluntary controls.

Part I Contamination of food
Section 1   If there is an escape of substances likely to create a hazard to human health through human consumption of food and the food may be in an area of the UK, Ministers may make an emergency order. The order may prohibit certain specified things including preparation, processing and movement of food. The first such order was made in June 1986 to prohibit the movement or slaughter of sheep from parts of Cumbria and North Wales following the escape of radioactivity from Chernobyl, USSR (see SI 1986/1027).

Part II Deposits in the sea

3.3

Part III Pesticides, etc.

Section 16   Control of Pesticides, etc.

Permits regulations to be made on all aspects of pesticide control
including specifying how much pesticide or pesticide residue may be
left in any crop, food or feeding stuff and provides for seizure and
destruction if the limits are exceeded.

Part IV General and supplemental

4 Food regulations and orders

4.1 A–Z food standards

Bread

Regulation: Bread and Flour Regulations 1995 (1995/3202)
Amendments: Food Labelling Regulations 1996 (1996/1499)
 Bread and Flour (Amendment) Regulations 1996 (1996/1501)

Definitions
Bread: a food of any size, shape or form which (a) is usually known as
 bread, and (b) consists of a dough made from flour and water, with or
 without other ingredients, which has been fermented by yeast or
 otherwise leavened and subsequently baked or partly baked; but does
 not include buns, bunloaves, chapatis, chollas, pitta bread, potato bread
 or bread specially prepared for coeliac sufferers.
Flour bleaching agent: any food additive primarily used to remove
 colour from flour.
Flour treatment agent: any food additive added to flour or dough to
 improve its baking quality other than an enzyme preparation.
Enzyme preparation: any food additive which consists of one or more
 enzymes with or without the addition of supplementary material to
 facilitate the storage, sale, standardisation, dilution or dissolution of the
 enzyme or enzymes.

General points
  1) Labelling restrictions: if the following terms are used as part of the
     name of a bread it must meet the specified conditions:
     a) wholemeal – all the flour used as an ingredient in the preparation
        of the bread is wholemeal;
     b) wheat germ – the bread has a minimum added processed wheat
        germ content of 10% (dry matter of bread).
  2) The Regulations came into effect on 1 January 1996. However, bread
     meeting the requirements of the previous controls (Bread and Flour
     Regulations 1994 as amended) may be sold until 1 July 1997.
  3) The Regulations do not apply to bread brought into Great Britain

4.1

from a European Economic Area (EEA) State in which it had been lawfully produced and sold or from an EU Member State in which it was in free circulation and lawfully sold. The bread must be suitably labelled to give the nature of the bread.

Permitted bleaching agents or treatment agents of bread and flour
Only agents listed in the following Table are permitted. Where a flour improver has been used as an ingredient of any bread its presence must be indicated in the ingredients list or, when the bread has no ingredients list, on a suitable label, ticket or notice.

Permitted ingredient	Types of flour and bread in which ingredient may be used	Maximum quantity, if any (mg/kg)
E 220 Sulphur dioxide E 223 Sodium metabisulphite	All flour intended for use in the manufacture of biscuits or pastry except wholemeal	(1)
E 300 L-Ascorbic acid	All flour except wholemeal All bread	200
920 L-Cysteine hydrochloride	a) All flour used in the manufacture of biscuits, except wholemeal or flour to which E220 or E223 has been added	300
	b) Other flour, except wholemeal All bread, except wholemeal	75
925 Chlorine	All flour intended for use in the manufacture of cakes, except wholemeal	2500
926 Chlorine dioxide	All flour except wholemeal All bread except wholemeal	30

Note: (1) The total quantity of these additives used must not exceed 200 mg/kg calculated as sulphur dioxide

Butter

Regulation: Spreadable Fats (Marketing Standards) Regulations 1995 (1995/3116)

See details given under 'Spreadable fats'.

Caseins and caseinates

Regulation: Caseins and Caseinates Regulations 1985 (1985/2026)
Amendments: Casein and Caseinates (Amendment) Regulations 1989 (1989/2321)
  Food Labelling Regulations 1996 (1996/1499)

Definitions
Casein: the principal protein constituent of milk, washed and dried, insoluble in water and obtained from skimmed milk by precipitation by

4.1

the addition of acid, or by microbial acidification, or by using rennet or by using other milk-coagulating enzymes, without prejudice to the possibility of prior ion exchange processes and concentration processes.

Caseinate: a product obtained by drying casein treated with neutralising agents.

Casein product: edible acid casein, edible casein or any edible caseinate.

Reserved descriptions

Edible acid casein: edible casein conforming to the standards below and precipitated using lactic acid (E 270), hydrochloric acid (507), sulphuric acid (513), citric acid (E 330), acetic acid (E 260), orthophosphoric acid (E 338), whey, bacterial cultures producing lactic acid.

Edible rennet casein: edible casein conforming to the standards below and precipitated using rennet, other milk-coagulating enzymes.

Edible caseinates: caseinates conforming to the standards below obtained from edible caseins using hydroxides/carbonates/phosphates/citrates of sodium/potassium/calcium/ammonium/magnesium of edible quality.

	Edible acid casein	Edible rennet casein	Edible caseinates
Moisture (max.)	10%	10%	8%
Milk protein (min.) (1)	90%	84%	
Milk protein (casein) (min.) (1)			88%
Milk fat (max.) (1)	2.25%	2%	2%
Titratable acidity (max.)	0.27 (2)		
pH value			6.0–8.0
Ash (P_2O_5 included)	2.5% max.	7.5% min.	
Anhydrous lactose (max.)	1%	1%	1%
Sediment content (max.) (3)	22.5 mg	22.5 mg	22.5 mg
Lead (max.)	1 mg/kg	1 mg/kg	1 mg/kg
Extraneous matter in 25 g (4)	nil	nil	nil
Odour	(5)	(5)	(6)
Appearance	(7)	(7)	(7)
Solubility			(8)

Notes:
(1) Calculated on the dried extract
(2) ml of decinormal sodium hydroxide solution per g
(3) Burnt particles in 25 g
(4) Such as wood or metal particles, hairs or insect fragments
(5) No foreign odours
(6) No more than very slight foreign flavours and odours
(7) Colour ranging from white to creamy white; the product must not contain any lumps that would not break up under slight pressure
(8) Almost entirely soluble in distilled water, except for the calcium caseinate

General points

1) Casein products must be labelled with the reserved description and,

4.1

in the case of caseinates (or mixtures of caseinates), an indication of the cation or cations.

2) Casein products sold as mixtures must be labelled as 'mixture of' followed by the relevant reserved descriptions (weight descending order).

3) Mixtures containing caseinates must be labelled with the protein content calculated on the dried extract expressed as % of the total weight of product as sold.

4) Any casein or caseinate used in the preparation of a casein product must be subjected to heat treatment at least equivalent to pasteurisation unless the casein product is itself subject to heat treatment during its preparation.

Chocolate products

Regulation: Cocoa and Chocolate Products Regulations 1976 (1976/541)
Amendments: Emulsifiers and Stabilisers in Food Regulations 1980 (1980/1833)
Miscellaneous Additives in Food Regulations 1980 (1980/1834)
Food Labelling Regulations 1980 (1980/1849)
Cocoa and Chocolate Products (Amendment) Regulations 1982 (1982/17)
Food Labelling Regulations 1984 (1984/1305)
Miscellaneous Additives in Food Regulations 1995 (1995/3187)
Food (Miscellaneous Revocations and Amendments) Regulations 1995 (1995/3267)
Food Labelling Regulations 1996 (1996/1499)

Definitions

Chocolate product: any product whose reserved description is given below but does not include any product specially prepared for diabetics or to which a slimming claim is lawfully applied and which has been specially prepared in connection with that claim by the addition of any ingredient other than an edible substance (as defined in the Regulations).

Chocolate: any product obtained from cocoa nib, cocoa mass, cocoa, fat-reduced cocoa (or any combination) and sucrose, with or without the addition of extracted cocoa butter and meeting the requirements in the table opposite.

Milk chocolate: any product obtained from cocoa nib, cocoa mass, cocoa, fat-reduced cocoa (or any combination) and sucrose, and from milk or milk solids, with or without the addition of extracted cocoa butter and meeting the requirements in the table opposite.

Reserved descriptions

	Minimum total dry cocoa solids (%)	Minimum dry non-fat cocoa solids (%)	Minimum permitted cocoa butter (%) (1)
Chocolate (2)	35	14	18
Quality chocolate (2)	43	14	26
Plain chocolate (2)	30	12	18
Chocolate vermicelli (3)	32	14	12
Chocolate flakes (3)	32	14	12
Gianduja nut chocolate (3, 4)	32	8	18
Couverture chocolate (2)	35	2.5	31
Dark couverture chocolate	35	16	31

	Minimum total dry cocoa solids (%)	Minimum dry non-fat cocoa solids (%)	Minimum milk solids (%)	Minimum milk fat (%)	Maximum sucrose (%) (5)	Minimum total fat (%)
Milk chocolate (2)						
(a)	25	2.5	14	3.5	55	25
(b)	20	2.5	20	5	55	25
Quality milk chocolate (2)	30	2.5	18	4.5	50	2.5
Milk chocolate vermicelli (3)	20	2.5	12	3	66	12
Milk chocolate flakes (3)	20	2.5	12	3	66	12
Gianduja nut milk chocolate (3,6)	25	2.5	10	3.5	55	25
Couverture milk chocolate (2)	25	2.5	14	3.5	55	31
White chocolate (2)		(7)	14	3.5	55	
Cream chocolate (2)	25	2.5	(8)	7	55	25
Skimmed milk chocolate (1)	25	2.5	(9)	(10)	55	25

Notes: All figures calculated excluding additional ingredients.
  (1) Permitted cocoa butter means fat in parts of cocoa beans having a level of
      unsaponifiable matter up to 0.35% max. determined using petroleum ether
      and an acidity of 1.75% max. expressed as oleic acid
  (2) Permitted additional ingredients:
      (a) 5–40% any edible substance in clearly visible and discrete pieces (except
          flour, starch or non-milk fat)
      (b) max. 30% any edible substance not in clearly visible and discrete pieces
          (except flour, starch, sucrose, dextrose, fructose, lactose, maltose or non-
          milk fat)
      (c) max. 5% vegetable fat (not derived from cocoa beans)
      (d) max. 10% partial coating or decoration (except for couverture and
          couverture milk chocolate)
      (e) max. 40% combinations of (a)–(d)
      (f) max. 30% combinations of (b)–(d)

4.1

(3) May also contain any flavouring substance which does not impart the flavour of chocolate or milk fat

(4) Contains 20–40% finely ground hazelnuts and may also contain:
  (a) whole or broken nuts up to a total nut content of 60%
  (b) milk or dry matter produced by partial or complete dehydration of whole milk or partially or fully skimmed milk to product with max. 5% dry milk solids including max. 1.25% butterfat

(5) Sucrose may be replaced by dextrose, fructose, lactose and/or maltose up to, in each case, 5% of product weight; if only dextrose is used it may be up to 20% of product weight

(6) Contains 15–40% finely ground hazelnuts and may also contain whole or broken nuts up to a total nut content of 60%

(7) Permitted cocoa butter min. 20% (see also (1))

(8) Dry non-fat milk solids 3–14%

(9) Dry non-fat milk solids min. 14%

(10) Less than 3.5% milk fat

Additional ingredients

1) All products may contain:
  a) any miscellaneous additive (subject to the Miscellaneous Food Additives Regulations 1995);
  b) any flavouring substance which does not impart the flavour of chocolate or milk fat.

2) There are certain additional detailed controls which allow for the use of additional edible substances at certain specified levels (see Regulations).

Labelling

1) There are detailed regulations governing statements to accompany the reserved description. The Regulations should be consulted for full details.

2) The presence of the following must not be indicated (except in an ingredients list) unless:
  a) for milk or milk products – the product is milk chocolate, couverture milk chocolate, white chocolate, cream chocolate or skimmed milk chocolate;
  b) for spirits – the amount of spirits is min. 1% of the chocolate product;
  c) for coffee – the amount of coffee solids (dry matter) is min. 1% of the chocolate product;
  d) for other edible substance not in discrete visible pieces – min. 5% of the chocolate product.

General points

1) The use of any of the above descriptions (or derivative or similar expression) may not be used unless the product conforms to the above standards except:

a) 'choc ice' or 'choc bar' may be used for an ice-cream with a coating resembling a chocolate product and containing min. 2.5% dry non-fat cocoa solids, and

4.1

b) 'choc roll' may be used for a swiss roll with the coating in (a), but, in both cases, the words must be accompanied by an appropriate designation to avoid confusion with any cocoa or chocolate product.

2) Reserved descriptions and labelling requirements are also applied to 'filled chocolate', 'X-filled Y' (or similar), 'a chocolate' and 'chocolates' (see Regulations).

3) Cocoa beans used in the preparation of a chocolate product must be sound, wholesome and in marketable condition.

4) The Regulations do not apply to products specially prepared for diabetics or to which a slimming claim is applied and which has thus been specially made by the addition of any ingredient other than an edible substance (as defined in the Regulations).

Cocoa products

Regulation: Cocoa and Chocolate Products Regulations 1976 (1976/541)
Amendments: Emulsifiers and Stabilisers in Food Regulations 1980 (1980/1833)
  Miscellaneous Additives in Food Regulations 1980 (1980/1834)
  Food Labelling Regulations 1980 (1980/1849)
  Cocoa and Chocolate Products (Amendment) Regulations 1982 (1982/17)
  Food Labelling Regulations 1984 (1984/1305)
  Miscellaneous Additives in Food Regulations 1995 (1995/3187)
  Food (Miscellaneous Revocations and Amendments) Regulations 1995 (1995/3267)
  Food Labelling Regulations 1996 (1996/1499)

Definitions
Cocoa product: any product listed in general point (1) below and any product whose reserved description is given below but does not include any product specially prepared for diabetics or to which a slimming claim is lawfully applied and which has been specially prepared in connection with that claim by the addition of any ingredient other than an edible substance (as defined in the Regulations).

Reserved descriptions

	Maximum water (%)	*Minimum permitted cocoa butter on dry matter (%) (1)*
Cocoa	9	20
Cocoa powder	9	20
Fat reduced cocoa	9	8
Fat reduced cocoa powder	9	8

4.1

	Minimum cocoa when mixed with sucrose (%) (2)
Sweetened cocoa	32
Sweetened cocoa powder	32
Drinking chocolate	25

	Minimum fat-reduced cocoa when mixed with sucrose (%) (2)
Sweetened fat-reduced cocoa	32
Sweetened fat-reduced cocoa powder	32
Fat-reduced drinking chocolate	25

Notes:
  (1) Permitted cocoa butter means fat in parts of cocoa beans having a level of unsaponifiable matter up to 0.5% max. determined using petroleum ether and an acidity of 1.75% max. expressed as oleic acid
  (2) Sucrose may be replaced by dextrose, fructose, lactose and/or maltose up to, in each case, 5% of product weight; if only dextrose is used it may be up to 20% of product weight

Additional ingredients
All products may contain:
  a) any miscellaneous additive (subject to the Miscellaneous Food Additives Regulations 1995);
  b) any flavouring substance which does not impart the flavour of chocolate or milk fat.

Labelling
There are very detailed regulations governing statements to accompany the reserved description. The Regulations should be consulted for details.

General points
  1) The Regulations also contain definitions and standards for the following cocoa products: cocoa bean, cocoa nib, cocoa dust, cocoa fines, cocoa mass, cocoa press cake, fat-reduced cocoa press cake, expeller cocoa press cake, cocoa butter, press cocoa butter, expeller cocoa butter, refined cocoa butter, cocoa fat.
  2) Cocoa beans used in the preparation of a cocoa product must be sound, wholesome and in marketable condition.
  3) The Regulations do not apply to products specially prepared for diabetics or to which a slimming claim is applied and which has thus been specially made by the addition of any ingredient other than an edible substance (as defined in the Regulations).

Coffee products

4.1

Regulation: Coffee and Coffee Products Regulations 1978 (1978/1420)
Amendments: Food Labelling Regulations 1980 (1980/1849)
    Coffee and Coffee Products (Amendment) Regulations 1982 (1982/254)
    Coffee and Coffee Products (Amendment) Regulations 1987 (1987/1986)
    Miscellaneous Additives in Food Regulations 1995 (1995/3187)
    Food Labelling Regulations 1996 (1996/1499)

Definitions
Coffee: the dried seed of the coffee plant (whether or not roasted and/or ground).
Coffee extract: the product in any concentration which contains the soluble and aromatic constituents of coffee, and is obtained by extraction from roasted coffee using only water as the medium of extraction (excluding any process of hydolysis involving the addition of an acid or base) and which may contain insoluble oils derived from coffee, traces of other insoluble substances not derived from coffee or from the water used for extraction.

Reserved descriptions
Coffee and chicory mixture: a mixture of only roasted coffee and chicory.
French coffee – coffee and chicory mixture: 51% minimum coffee with no matter other than roasted coffee and chicory.
Coffee with fig flavouring (Viennese coffee): 85% minimum coffee with no matter other than roasted coffee and fig.

	Dry matter (%)	When present, maximum added sugar (%)
Dried coffee extract (dried extract of coffee, soluble coffee, instant coffee)	min. 95 (2)	
Coffee extract paste	70–85 (2)	
Liquid coffee extract	15–55 (2)	12
Dried chicory extract (soluble chicory, instant chicory)	min. 95 (3)	
Chicory extract paste	70–85 (3)	
Liquid chicory extract	25–55 (3)	35
Dried coffee and chicory extract (1)	min. 95 (4)	
Coffee and chicory paste (1)	70–85 (4)	
Liquid coffee and chicory extract (1)	15–55 (4)	25
Chicory and coffee essence	(6)	(7)
Dried coffee and fig extract (1)	min. 95 (5)	
Coffee and fig paste (1)	70–85 (5)	
Liquid coffee and fig extract (1)	15–55 (5)	25

See overleaf for Notes

4.1

Notes:
(1) Name of ingredient of which higher proportion used in manufacture to appear first
(2) Coffee-based dry matter content
(3) Chicory-based dry matter content
(4) Coffee and chicory-based dry matter content
(5) Coffee and fig-based dry matter content
(6) Min. 20% chicory-based dry matter and min. 5% coffee-based dry matter
(7) May contain added sugar

Additional labelling provisions
Products should, for retail sale, be labelled with:
a) a reserved description;
b) the word 'decaffeinated' if it has been subject to a decaffeination process and in which the residual anhydrous caffeine content does not exceed (based on coffee-based dry matter)
   i) 0.10% for coffee, coffee and chicory mixtures, coffee with fig flavouring,
   ii) 0.30% for coffee extracts and extracts of blends;
c) in the case of products above with added sugar,
   i) 'roasted with sugar' if the raw material has been roasted with sugar,
   ii) 'with sugar', 'preserved with sugar' or 'with added sugar' if sugar added after roasting,
   and in both cases the word 'sugar' replaced by the reserved description of the sugar used;
d) in the case of coffee/chicory extract paste and liquid coffee/chicory extract (or mixtures with these), a declaration of the minimum % coffee/chicory-based dry matter;
e) in the case of liquid coffee extract with min. 25% coffee-based dry matter or liquid chicory extract with min. 45% chicory-based dry matter, the word 'concentrated' may be used.

Permitted ingredients
1) Any miscellaneous additive (subject to the Miscellaneous Food Additives Regulations 1995).
2) Decaffeination agent: may be added to decaffeinated coffee which satisfies Notes (1) or (2) in the table above as long as permitted by Food Safety Act 1990.
3) Added sugar products: permitted only where indicated in above table.

General points
1) All raw materials should be sound, wholesome and in marketable condition.

2) The word 'coffee' may be used to describe a beverage prepared from the following: dried coffee extract, coffee extract paste or liquid coffee extract; and 'dandelion coffee' may be used to describe a product consisting of or made from roasted dandelion root.

4.1

Fish (preserved)

A Sardines

Regulations: Preserved Sardines (Marketing Standards) Regulations 1990 (1990/1084) EEC Council Regulation 2136/89

Only products meeting certain specified requirements may be marketed using specified trade descriptions. Included are the following points (see Council Regulation 2136/89 for full details):

1) Only fish of the species *Sardinia pilchardus* Walbaum may be used (with an exception concerning products using homogenised sardine flesh). The fish should be appropriately trimmed of the heads, gills, caudal fin and internal organs (other than the ova, milk and kidneys) and, as appropriate, backbone and skin. However, provision is made for certain preparations using homogenised sardine flesh (minimum 25% sardines) to contain the flesh of other fish.

2) They must be pre-packaged with an appropriate covering medium. The following are listed and shall be indicated in the trade description: olive oil (not to be mixed with other oils); other refined vegetable oils, including olive-residue oil used singly or in mixtures; tomato sauce; natural juice (liquid exuding from the fish during cooking), saline solution or water; marinade, with or without wine.

   Other covering media may be used on condition that they are clearly distinguished from those in the above list.

3) Preserved sardines may be marketed in the following presentations (Regulation 2136/89 gives more specific details relating to the cuts to be used):
   a) sardines;
   b) sardines without bones;
   c) sardines without skin or bones;
   d) sardine fillets;
   e) sardine trunks;
   f) other presentations – these may be used on condition that they are clearly distinguished from those above.

4) The pack should be sterilised by appropriate treatment.

5) After sterilisation, the products must meet the following conditions:
   a) For the presentations (a) to (e) in (3), the sardines must: be reasonably uniform in size and arranged in an orderly manner in the container; be readily separable from each other; present no significant breaks in the abdominal wall; present no breaks or

4.1

tears in the flesh; present no yellowing of tissues (with the exception of slight traces); comprise flesh of normal consistency (the flesh must not be excessively fibrous, soft or spongy); comprise flesh of a light or pinkish colour, with no reddening round the backbone with the exception of slight traces.

b) The covering medium must have the appropriate characteristic colour and consistency. In the case of an oil medium, the maximum aqueous exudate is 8% of net weight.

c) The product must have the appropriate characteristic odour and flavour and must be free of any disagreeable odour or taste.

d) Be free of foreign bodies.

e) For products with bones, the backbone must be readily separable.

f) For products without skin and bones, there should be no significant residues of these.

6) The pack may not present external oxidation or deformation affecting good commercial presentation.

7) The ratio of the weight of the sardines (for the presentations (a) to (e) in (3)) in the container after sterilisation and the net weight depends upon covering media and shall be not less than:

olive oil; other refined vegetable oils; natural juice, saline solution or water; marinade: 70%

tomato sauce: 65%

other media: 50%

For other presentations (f) in (3), the ratio must be not less than 35%.

B   Tuna and bonito

Regulation: Preserved Tuna and Bonito (Marketing Standards) Regulations 1994 (1994/2127)
EEC Council Regulation 1536/92

Only products meeting certain specified requirements may be marketed using specified trade descriptions. Included are the following points (see Council Regulation 1536/92 for full details):

1) The products must be prepared exclusively from fish of one of the following species:

a) For preserved tuna: albacore or longfinned tuna (*Thunnus alahunga*), yellowfin tuna (*Thunnus (neothunnus) albacores*), bluefin tuna (*Thunnus (parathunnus) obesus*), other species of the genus *Thunnus*, skipjack or stripe-bellied tuna (*Euthynnus (katsuwonus) pelamis*).

b) For preserved bonito: Atlantic bonito (*Sarda sarda*), Pacific bonito (*Sarda chiliensis*), Oriental bonito (*Sarda orientalis*), other species of the genus *Sarda*, Atlantic little tuna (*Euthynnus alleteratus*), Eastern little tuna (*Euthynnus affinis*), black skipjack (*Euthynnus lineatus*),

4.1

other species of the genus *Euthynnus* (except *Euthynnus (katsuwonus) pelamis*), frigate mackerel (*Auxis thazard*), *Auxis rochei*. However, provision is made for certain culinary preparations using tuna and/or bonito flesh (minimum 25% tuna and/or bonito) which does not display muscular structure to contain the flesh of other fish.

2) Products marketed with the following words, relating to the presentation, as part of their trade description shall comply with specified requirements (see Regulation 1536/92 for full details) (other forms or culinary preparations are permitted but shall be clearly identified in the trade description):
   a) solid (includes requirement of maximum 18% flakes except when canned raw when flakes are prohibited);
   b) chunks (includes requirement of maximum 30% flakes);
   c) fillets;
   d) flakes;
   e) grated/shredded tuna.

3) Products marketed with the following words, relating to the covering medium, as part of their trade description shall comply with specified requirements (where any other covering medium is used, it shall be identified clearly and explicitly using its usual trade name):
   a) 'in olive oil' – for products using only olive oil (mixtures with other oils not permitted);
   b) 'natural' – for products using natural juice (i.e. the liquid exuding from the fish during cooking), a saline solution or water, possibly with the addition of herbs, spices or natural flavourings;
   c) 'in vegetable oil' – for products using refined vegetable oils (singly or mixtures).

4) Trade descriptions shall contain the following:
   a) for products defined in (2) (a) to (e) above – the type of fish, the presentation (although in the case of 'solid' it may be omitted), and the covering medium;
   b) for other presentations – the type of fish (tuna or bonito), and the precise nature of the culinary preparation.

5) The words 'tuna' and 'bonito' must not be associated in the trade description (except where there has been traditional national usage).

6) The word 'natural' may only be used for products with presentations defined in (2) (a) to (c) above and for the covering medium described in (3) (b) above.

7) The ratio between fish weight after sterilisation and the net weight shall be as follows:
   a) for products defined in (2) (a) to (e) above:
      i) if the covering medium is 'natural' 70%,
      ii) for other covering media 65%,
   b) for other presentations 25%.

4.1 Flour

Regulation: Bread and Flour Regulations 1995 (1995/3202)
Amendments: Food Labelling Regulations 1996 (1996/1499)
  Bread and Flour (Amendment) Regulations 1996 (1996/1501)

Definitions
Flour: the product derived from, or separated during, the milling or grinding of cleaned cereal, whether or not the cereal has been malted or subjected to any other process, and includes meal, but does not include other cereal products, such as separated cereal bran, separated cereal germ, semolina or grits.

Composition of wheat flour
1) All wheat flour, except wholemeal, self-raising flour (with a calcium content of min. 0.2%), and wheat malt flour, shall contain:
   – calcium carbonate, between 235 and 390 mg/100 g of flour.
2) All wheat flour shall contain (per 100 g of flour):
   – iron, minimum 1.65 mg;
   – thiamin (vitamin B_1) minimum 0.24 mg;
   – nicotinic acid or nicotinamide, mininimum 1.60 mg.
   For wholemeal, the substances should be naturally present and not added.
3) These standards do not apply to flour which is for use in the manufacture of communion wafers, matzos, gluten, starch or any concentrated preparation used in facilitating the addition to flour of the above substances.

Permitted flour bleaching agents or flour treatment agents
See table of 'Permitted bleaching agents or treatment agents of bread and flour' given under 'Bread'.

General points
1) The Regulations came into effect on the 1 January 1996. However, flour meeting the requirements of the previous controls (Bread and Flour Regulations 1994, as amended) may be sold until 1 July 1997.
2) The Regulations do not apply to flour brought into Great Britain from an EC Member State in which it had been lawfully produced and sold or in which it was in free circulation and lawfully sold. The flour must be suitably labelled to give the nature of the flour.

Fruit juices and nectars

Regulation: Fruit Juices and Fruit Nectars Regulations 1977 (1977/927)
Amendments: Lead in Food Regulations 1979 (1979/1254)
  Food Labelling Regulations 1980 (1980/1849)
  Fruit Juices and Fruit Nectars (Amendment) Regulations 1982 (1982/1311)

Fruit Juices and Fruit Nectars (England, Wales and Scotland) (Amendment) Regulations 1991 (1991/1284)

4.1

Fruit Juices and Fruit Nectars (England, Wales and Scotland) (Amendment) Regulations 1995 (1995/236)

Miscellaneous Food Additives Regulations 1995 (1995/3187)

Food (Miscellaneous Revocations and Amendments) Regulations 1995 (1995/3267)

Food Labelling Regulations 1996 (1996/1499)

Definitions

Fruit juice: either
   a) the food consisting of fermentable but unfermented product which is obtained from (i) fruit by mechanical processes and has the characteristic colour, aroma and flavour of juice of the fruit from which it is obtained, or (ii) concentrated fruit juice by the addition of water and has the same organoleptic and analytical characteristics of fruit juice obtained from fruit of the same kind by mechanical processes, or (iii) fruit (other than apricots, citrus fruits, grapes, peaches, pears or pineapples) by diffusion processes and is intended to be used in the preparation of concentrated fruit juice; or
   b) fruit purée where the nature of the fruit from which the juice is to be obtained is such that it is impossible to extract the juice without the pulp.

Fruit purée: the fermentable but unfermented product obtained by sieving the entire edible part of whole or peeled fruit.

Fruit nectar: the food consisting of the fermentable but unfermented product obtained by the addition of water and sugar to fruit juice, concentrated fruit juice, fruit purée or concentrated fruit purée (or a mixture of these) and satisfies the following table except however:
   a) maximum added sugar 20% (finished product, dry matter basis);
   b) maximum added honey 20% (finished product);
   c) where note (1) in the table applies;
   d) where the naturally high sugar content of apricots or any fruit in parts 2 and 3 of the table (or any mixture of these) so warrants, they may be used to manufacture the product without the addition of sugar or honey.

Fruit from which the product is obtained	*Minimum quantity of acid expressed as tartaric acid* (g/l finished product)	*Minimum quantity of juice and/or purée* (% wt (finished product))
1. Apricots	3 (1)	40
Bilberries	4	40
Blackberries	6	40
Blackcurrants	8	25
Cherries (other than sour cherries)	6 (1)	40
Cranberries	9	30

37

4.1 (*continued*)

Fruit from which the product is obtained	Minimum quantity of acid expressed as tartaric acid (g/l finished product)	Minimum quantity of juice and/or purée (% wt (finished product)
Elderberries	7	50
Gooseberries	9	30
Lemons	–	25
Limes	–	25
Mulberries	6	40
Passion fruit (*Passiflora edulis*)	8	25
Plums	6	30
Quetsches	6	30
Quinces	7	50
Quito Naranjillos (*Solanum quitoense*)	5	25
Raspberries	7	40
Redcurrants	8	25
Rose hips (fruits of the species *Rosa*)	8	40
Rowanberries	8	30
Sallowthorn berries	9	25
Sloes	8	30
Sour cherries	8	35
Strawberries	5	40
Whitecurrants	8 (1)	25
Any other fruit with acid juice unpalatable in the natural state	–	25
2. Azeroles (Neopolitan medlars)	–	25
Bananas	–	25
Bullock's heart (Custard apple) (*Annona reticulata*)	–	25
Cashew fruits	–	25
Guavas	–	25
Lychees	–	25
Mangoes	–	35
Papayas	–	25
Pomegranates	–	25
Soursop (*Annona muricata*)	–	25
Spanish plums (*Spondia purpurea*)	–	25
Sugar apples	–	25
Umbu (*Spondias tuberosa aroda*)	–	30
Any other low acid, pulpy or highly flavoured fruit with juice unpalatable in the natural state	–	25
3. Apples	3 (1)	50
Peaches	3 (1)	45
Pears	3 (1)	50
Pineapples	4	40
Citrus fruits other than lemons and limes	5	50
Any other fruit with juice palatable in the natural state	–	50

Note: (1) Does not apply when product obtained exclusively from fruit purée and/or concentrated fruit purée

Permitted additional ingredients

When for sale for consumption, only the following additional ingredients
are permitted in fruit juice, concentrated or dried fruit juice or fruit
nectar (or concentrated fruit juice or fruit juice intended for the
preparation of these).

1) Any miscellaneous additive (subject to the Miscellaneous Food
   Additives Regulations 1995) except that no apple juice, grape juice,
   pineapple juice or concentrated pineapple juice shall contain both
   added sugar and added acid.

2) Concentrated fruit juice, fruit juice derived from concentrated fruit
   juice and dried fruit juice may contain volatile components of the
   same fruit, which need not be included in the ingredients list.

3) Fruit juice and concentrated or dried fruit juice may contain, after
   dilution or reconstitution if appropriate, sugar (types specified in the
   Regulations) as follows:
   - grapes/pears, not permitted;
   - apples, up to 40 g/l;
   - bergamots, blackcurrants, lemons, limes, redcurrants or white
     currants, up to 200 g/l;
   - other fruits, up to 100 g/l.

   For concentrated orange juice not prepackaged and not for sale to the
   ultimate consumer, added sugar is permitted up to 15 g/l (calculated
   as dry matter and after dilution to preconcentrated strength).

4) The following fruit nectars may contain lemon juice in total or partial
   replacement of citric acid up to a maximum of 5 g/l:
   a) apple nectar obtained exclusively from apple purée or
      concentrated purée (or mixture of these)
   b) peach nectar obtained exclusively from peach purée or
      concentrated purée (or mixture of these)
   c) pear nectar obtained exclusively from pear purée or concentrated
      purée (or mixture of these)
   d) any mixture of (a), (b) and (c)

Specific labelling requirements

1) Product to be labelled 'juice', 'concentrated juice', 'dried juice' or
   'nectar' as appropriate and preceded by the name (or names in
   weight descending order, after reconstitution if necessary) of the
   fruit(s) used (but not including any lemon juice added as acid to
   nectar).

2) The word 'dried' may be replaced by 'powdered' and may be
   accompanied or replaced by 'freeze-dried' (or other word descriptive
   of the specific process).

3) Where sugar has been added (in accordance with (5) above) and
   exceeds 15 g/l, the product shall be labelled 'sweetened' and there

4.1

should be a declaration of the maximum added sugar (which must not exceed the actual sugar content by more than 15%).

4) Where fruit nectar is made from fruit purée or concentrated fruit purée, product to be labelled as 'contains fruit pulp'.

5) Where fruit juice or fruit nectar is made from concentrated fruit juice (or purée), product to be labelled as 'made with concentrated … juice (purée)' (name of fruit to be added).

6) Fruit juice or fruit nectar which contains carbon dioxide at a level which exceeds 2 g/l must be labelled 'carbonated'.

7) Fruit nectar shall be marked with a statement 'fruit content $x\%$ minimum'.

8) Concentrated or dried fruit juice shall be marked with the quantity of water required for reconstitution.

General points

1) Most of the above standards do not apply to any concentrated fruit juice specially prepared for infants and children and labelled as such (see Regulations).

2) Concentrated fruit juice must have been reduced in volume by at least 50%.

3) For citrus fruits, the juice must only be derived from the endocarp (although for lime juice the minimum possible of constituents from the outer parts of the fruit is permitted).

4) The preparation of concentrated or dried fruit juice must not involve the application of direct heat.

Honey

Regulation: Honey Regulations 1976 (1976/1832)
Amendment: Food Labelling Regulations 1980 (1980/1849)
 Food Labelling Regulations 1996 (1996/1499)

Definitions

Honey: the fluid, viscous or crystallised food which is produced by honeybees from the nectar of blossoms, or from secretions of, or found on, living parts of plants other than blossoms, which honeybees collect, transform, combine with substances of their own and store and leave to mature in honeycombs.

Blossom honey: honey produced wholly or mainly from the nectar of blossoms.

Honeydew honey: honey, the colour of which is light brown, greenish brown, black or any intermediate colour, produced wholly or mainly from secretions of or found on living parts of plants other than blossoms.

Pressed honey: honey obtained by pressing broodless honeycombs with or without the application of moderate heat.

Reserved descriptions

4.1

	Minimum apparent reducing sugar content (%) (1)	Maximum moisture content (%) (2)	Maximum apparent sucrose content	Maximum water-insoluble solids (%)	Maximum ash content (%)
Honeydew honey	60	21	10	0.1	1
Blended honeydew honey and blossom honey	60	21	10	0.1	1
Heather honey	65	23	5	0.1	0.6
Clover honey	65	23	5	0.1	0.6
Acacia honey	65	21	10	0.1	0.6
Lavendar honey	65	21	10	0.1	0.6
Banksia menzieii honey	65	21	10	0.1	0.6
Pressed honey	65	21	5	0.5	0.6
Honey	65	21	5	0.1	0.6

Notes:
(1) Calculated as invert sugar.
(2) Moisture content may be above specified standard but not more than 25% if such moisture content is the result of natural conditions of production.

Bakers' or industrial honey

Honey with one or more of the following conditions:
1) fails above moisture restrictions,
2 has any foreign tastes or odours,
3) has begun to ferment or effervesce,
4) has been heated to such an extent that its natural enzymes have been destroyed or made inactive,
5) is citrus honey (or any other honey with a naturally low enzyme content) with a diastase activity of less than 3 (determined as specified in the Regulations),
6) is honey not included in (5) with a diastase activity of less than 4,
7) has a hydroxymethylfurfural content of more than 80 mg/kg.

General points
1) No person shall add to honey (intended for sale) any substance other than honey.
2) Honey (sold, consigned or delivered) shall be free, as far as practicable, from mould, insects, insect debris, brood or any other organic substance foreign to the composition of honey. Honey used in the preparation of another food shall also be free of these substances.
3) Honey shall have an acidity of not more than 40 mEq acid/kg. No person may sell honey with an artificially changed acidity.

4.1

4) Reference to origin:
   a) if a type of blossom or plant is indicated, the honey must be wholly or mainly from that source,
   b) if a regional, topographical or territorial origin is indicated, the honey must originate wholly in that place.
5) The compositional standards in the table do not apply to any comb contained in comb honey or the comb in chunk honey.

Infant formula and follow-on formula

Regulation: Infant Formula and Follow-on Formula Regulations 1995 (1995/77)

Definitions
Follow-on formula: a food intended for particular nutritional use by infants in good health who are aged over 4 months, and constituting the principal liquid element in a progressively diversified diet.
Infant formula: a food intended for particular nutritional use by infants in good health during the first 4 to 6 months of life, and satisfying by itself the nutritional requirements of such infants.
Infant: a child under the age of 12 months.
Young children: children between 1 and 3 years.

Essential composition
Note: In the Regulations, data are given based on units expressed in both kJ and kcal. Only those using kJ are given below.
  1) The following are specified for infant formulae and follow-on formulae when reconstituted as instructed by the manufacturer (all values refer to the product ready for use) but the table should be read in conjunction with the notes to the table and points (2) to (4) below:

	Infant formulae		*Follow-on formulae*	
	Minimum	*Maximum*	*Minimum*	*Maximum*
Energy (kJ)	250	315	250	335
Protein (g/100kJ) (1)	–	–	0.5	1
a) Infant formulae from unmodified cows' milk proteins	0.56	0.7	–	–
b) Infant formulae from modified cows' milk proteins	0.45	0.7	–	–

(continued)

4.1

	Infant formulae		Follow-on formulae	
	Minimum	Maximum	Minimum	Maximum
c) Infant formulae from soya protein isolates (alone or mixed with cows' milk proteins)	0.56	0.7	–	–
Lipids (g/100kJ)	0.8	1.5	0.8	1.5
a) Lauric acid	–	15% of total fat content	–	15% of total fat content
b) Myristic acid	–	15% of total fat content	–	15% of total fat content
c) Linoleic acid (in the form glycerides (linoleates)) (mg/100kJ)	70	285	70 (2)	
Carbohydrates (g/100kJ)	1.7	3.4	1.7	3.4
a) Lactose	0.85 (3)	–	0.45 (3)	
b) For infant formulae, sucrose; for follow-on formulae, sucrose, fructose, honey (separately or as a whole)	–	20% of total carbohydrate content		20% of total carbohydrate content
c) precooked starch and/or gelatinised starch (g/100 ml)	–	2/100, and 30% of total carbohydrate content		
Minerals			(4)	
a) Sodium (mg/100 kJ)	5	14		
b) Potassium (mg/100 kJ)	15	35		
c) Chloride (mg/100 kJ)	12	29		
d) Calcium (mg/100 kJ)	12	–		
e) Phosphorus (mg/100 kJ)	6	22		
Calcium/phosphorus ratio	1.2	2.0	–	2.0
f) Magnesium (mg/100 kJ)	1.2	3.6		
g) Iron (mg/100 kJ) (5)	0.12 (6) or 0.25 (7)	0.36 (6) or 0.5 (7)	0.25	0.5
h) Zinc (mg/100kJ)	0.12 (6) or 0.18 (7)	0.36 (6) or 0.6 (7)	0.12 (6) or 0.18 (7)	–
i) Copper (µg/100kJ)	4.8	19		
j) Iodine (µg/100kJ)	1.2	–	1.2	
Vitamins				
a) Vitamin A (µg RE/100 kJ)(8)	14	43	14	43
b) Vitamin D (µg/100 kJ)(9)	0.25	0.65	0.25	0.75
c) Thiamin (µg/100 kJ)	10	–	–	–
d) Riboflavin (µg/100 kJ)	14	–	–	–
e) Nicotinamide (µg NE/100 kJ)(10)	60	–	–	–
f) Pantothenic acid (µg/100 kJ)	70	–	–	–
g) Vitamin B_6 (µg/100 kJ)	9	–	–	–
h) Biotin (µg/100 kJ)	0.4	–	–	–
i) Folic acid (µg/100 kJ)	1	–	–	–

4.1

(continued)

	Infant formulae		Follow-on formulae	
	Minimum	Maximum	Minimum	Maximum
j) Vitamin B_{12} (μg/100 kJ)	0.025	–	–	–
k) Vitamin C (mg/100 kJ)	1.9	–	1.9	–
l) Vitamin K (μg/100 kJ)	1	–	–	–
m) Vitamin E				
(mg α-TE/100 kJ)(11)	0.5 (12)	–	0.5 (12)	–

Notes:

(1) Protein content is calculated as nitrogen content × 6.36 (for cows' milk proteins) or × 6.25 for soya protein isolates

(2) This limit applies only to follow-on formulae containing vegetable oils

(3) This provision does not apply to formulae in which soya proteins represent more than 50% of the total protein content

(4) Except where specified in the table, mineral concentrations for follow-on formulae should be at least equal to those normally found in cows' milk, reduced (if appropriate) in the same ratio as the protein concentration of the follow-on formulae to that of cows' milk (reference data for cows' milk are given in the Regulations)

(5) For infant formulae, the limit is applicable to formulae with added iron

(6) When manufactured from cows' milk proteins

(7) When manufactured from soya protein isolates (alone or mixed with cows' milk proteins)

(8) RE, all-*trans* retinol equivalent

(9) In the form of cholecalciferol, of which 10 μg = 4090 i.u. of vitamin D

(10) NE, niacin equivalent = mg nicotinic acid + mg tryptophan/60.

(11) α-TE, o-α-tocopherol equivalent.

(12) Per g of polyunsaturated acids expressed as linoleic acid but in no case less than 0.1 mg per 100 available kJ

2) Proteins.

   a) For infant formulae manufactured from unmodified cows' milk proteins: the chemical index of the proteins present shall be at least 80% of that of breast milk (see Regulations for definition of chemical index and for reference data for breast milk).

   b) For infant formulae manufactured from modified cows' milk proteins: for an equal energy value, the formula must contain an available quantity of each essential and semi-essential amino acid at least equal to that contained in breast milk (see Regulations for reference data for breast milk).

   c) For infant formulae from soya protein isolates (alone or mixed with cows' milk proteins): only soya protein isolates may be used; the chemical index shall be at least equal to 80% of that of breast milk (see Regulations for reference data for breast milk); for an equal energy value, the formula must contain an available quantity of methionine at least equal to that contained in breast milk; the L-carnitine content shall be at least 1.8 μmol/100 kJ.

d) For any follow-on formulae: the chemical index shall be at least equal to 80% of that of casein (see Regulations for definition of chemical index and reference data for casein); if manufactured from soya proteins, alone or mixed with cows' milk proteins, only soya protein isolates may be used.

e) In all cases (a)–(d), the addition of amino acids is permitted solely for improving the nutritional value of the proteins (and only in the necessary proportion).

3) Lipids: for both infant formulae and follow-on formulae, the use of the following is prohibited – sesame seed oil, cotton seed oil, fats containing more than 8% *trans* isomers of fatty acids.

4) Carbohydrates:

   a) For infant formulae, only the following carbohydrates may be used: lactose, maltose, sucrose, maltodextrins, glucose syrup or dried glucose syrup, precooked starch and gelatinised starch (both starches to be naturally free of gluten).

   b) For follow-on formulae, the use of ingredients containing gluten is prohibited.

5) The Regulations list those nutritional ingredients which may be used to meet the specified requirements for vitamins, minerals, amino acids, other nitrogen compounds and other substances (see Regulations).

6) In addition to the ingredients listed in the above requirements and table, other food ingredients may be used if their suitability for the particular nutritional use by infants from birth (for infant formulae) or by infants over 4 months (for follow-on formulae) has been established by generally accepted scientific data.

Labelling

1) Infant formulae and follow-on formulae must be labelled with the following:

   a) if protein source not entirely cows' milk, the name 'infant formula' or 'follow-on formula' as appropriate;

   b) if protein source entirely cows' milk, the name 'infant milk' or 'follow-on milk' as appropriate;

   c) available energy (in kJ and kcal), and, proteins, lipids and carbohydrates (per 100 ml ready for use);

   d) the average quantity of each mineral and vitamin mentioned in table above and, where appropriate, choline, inositol and carnitine (all per 100 ml ready for use);

   e) instructions for appropriate preparation and a warning against the health hazards of inappropriate preparation.

2) In addition to items in (1), infant formulae must be labelled with the following:

4.1

   a) a statement to the effect that the product is suitable for particular nutritional use by infants from birth when they are not breast-fed;

   b) if product does not contain added iron, a statement to the effect that the product does not contain the total iron requirements recommended for an infant over the age of 4 months, and that these should be made up from additional sources;

   c) the words 'Important Notice' immediately followed by:
- a statement concerning the superiority of breast-feeding, and
- a statement recommending that the product be used only on the advice of an independent person qualified in medicine, nutrition or pharmacy or having a professional qualification in maternal or child-care.

3) Infant formulae must not be labelled with the following:

   a) any picture of an infant,

   b) any other picture or text which may idealise the use of the product.

But it may include graphic representations for easy identification of the product or for illustrating methods of preparation.

4) Infant formulae may make the following claims when the specified conditions are met:

   a) 'Adapted protein': the protein content is lower than $0.6 \text{ g}/100 \text{ kJ}$ and the whey protein/casein ratio is not less than 1.0.

   b) 'Low sodium': sodium content lower than $9 \text{ mg}/100 \text{ kJ}$.

   c) 'Sucrose free': no sucrose present.

   d) 'Lactose only': lactose is the only carbohydrate present.

   e) 'Lactose free': no lactose present.

   f) 'Iron enriched': iron is added.

5) In addition to items in (1), follow-on formulae must be labelled with the following: a statement to the effect that the product is suitable for particular nutritional use by infants over the age of 4 months, that it should form only a part of a diversified diet and that it is not to be used as a substitute for breast milk during the first four months of life.

6) The labelling of any infant formulae and follow-on formulae must:

   a) be designed to provide the necessary information about the appropriate use of the product so as not to discourage breast-feeding;

   b) not contain the terms 'humanized', 'maternalized' or any similar term suggesting that the product is equivalent or superior to breast milk;

and the term 'adapted' may only be used as provided in (4) (a) above.

General points

1) The Regulations contain additional restrictions on the advertising and promotion of infant formulae and follow-on formulae and on the

provision of informational and educational material regarding infant and child feeding for pregnant women and mothers of infants and young children (see Regulations).

4.1

2) Exports must comply with the specified requirements or the requirements of the Codex Standard for Infant Formula and for Follow-up Formula or the requirements of the importing country. They must be labelled in one or more appropriate languages and in a way to avoid confusion between infant formulae and follow-on formulae.

3) If an infant formula or follow-on formula is not ready to use at the time of sale, only water shall be needed to make it ready.

Jam (and similar products)

Regulation: Jam and Similar Products Regulations 1981 (1981/1063)
Amendments: Food Labelling (Amendment) Regulations 1982 (1982/1700)
  Sweeteners in Food Regulations 1983 (1983/1211)
  Sweeteners in Food (Amendment) Regulations 1988 (1988/2112)
  Preservatives in Food Regulations 1989 (1989/533)
  Jam and Similar Products (Amendment) Regulations 1990 (1990/2085)
  Sweeteners in Food Regulations 1995 (1995/3123)
  Colours in Food Regulations 1995 (1995/3124)
  Miscellaneous Food Additives Regulations 1995 (1995/3187)
  Food Labelling Regulations 1996 (1996/1499)

Note: In these Regulations, fruit includes carrots, ginger, rhubarb and sweet potatoes.

Reserved descriptions
Jam/extra jam/marmalade: a mixture, brought to a suitable gelled consistency, of sweetening agents and fruit (see Note (1) in table overleaf), and complying with the following tables.
Jelly/extra jelly/UK standard jelly: an appropriately gelled mixture of sweetening agents and fruit (see note (1) in table overleaf), and complying with the following tables.
Sweetened chestnut purée: a mixture, brought to a suitable gelled consistency, of sweetening agents and puréed chestnuts, and complying with the following tables.

4.1

Minimum quantity of fruit (1) used for 1 kg of product (if mixed fruits are used, figures are reduced in proportion to relative use of fruit):

	Extra jam or extra jelly (g) (2)	Jam or jelly or UK standard jelly (g)	Marmalade (g)	Sweetened chestnut purée (g)
Passion fruit	80	60		
Cashew apples	230	160		
Ginger	250	150		
Blackcurrants, rosehips, quinces	350	250		
Citrus fruit			200 (3)	
Puréed chestnuts				380
Other fruit	450	350		

Notes:
  (1) Fruit defined as follows:
      for extra jam – fruit pulp (but see also notes (2) and (4))
      for jam – fruit pulp and/or fruit purée
      for jelly and extra jelly – fruit juice and/or aqueous extract of fruit (but see also note (2))
      for UK standard jelly – fruit juice and/or aqueous extract of fruit derived from the quantity of fruit, fruit pulp and/or fruit purée stated in the table
      for marmalade: fruit pulp, fruit purée, fruit juice, fruit peel or aqueous extract of fruit, in every case obtained from citrus fruit
  (2) Extra jam or extra jelly, if made from a mixture of fruit, must not include pulp of the following: apples, pears, clingstone plums, melons, watermelons, grapes, pumpkins, cucumbers, tomatoes.
  (3) Minimum 75 g of endocarp.
  (4) For extra jam, where only fruit is rosehips, fruit may be fruit pulp and/or fruit purée.

X curd or Y flavour curd
(Where X is a named fruit, 'mixed fruit' or a specified number of fruits; and Y is a named fruit or 'mixed fruit')
X curd: an emulsion of edible fat or oil (or both), sugar, whole egg or egg yolk (or both), and any combination of fruit, fruit pulp, fruit purée, fruit juice, aqueous extract of fruit or essential oils of fruit, with or without other ingredients, and meeting the following standards.
Y flavour curd: an emulsion of edible fat or oil (or both), sugar, whole egg or egg yolk (or both), and flavouring material with or without other ingredients, and meeting the following standards.

  a) Fat or oil – min. 40 g/kg finished product.
  b) Egg yolk solids (derived from whole egg and egg yolk) – min. 6.5 g/kg finished product.
  c) For X curd – sufficient fruit (as listed above) or essential oil of fruit to characterise the finished product.
  d) For Y flavour curd – sufficient flavouring material to characterise the product.

Lemon cheese
A food conforming to the requirements above for X curd appropriate for
lemon curd.

4.1

Mincemeat
A mixture of sweetening agents, vine fruits, citrus peel, suet or equivalent
fat and vinegar or acetic acid, with or without other ingredients, such that
the following are found in each kg of finished product:
  a) vine fruits and citrus peel: min. 300 g,
  b) vine fruits – min. 200 g,
  c) suet (or equivalent fat): min. 25 g.

Permitted ingredients
  1) Any miscellaneous additive permitted by the Miscellaneous Food
     Additives Regulations.
  2) In addition to those ingredients specified in the reserved descriptions
     above, the following may contain the ingredients indicated and their
     use must be indicated in the name of the food.
     a) All products specified above: any edible ingredient other than
        citrus fruit juice and additives mentioned in the rest of this item.
     b) Extra jam, jam: citrus fruit juice.
     c) Extra jam, jam, extra jelly and jelly (in each made from quinces):
        citrus peel, leaves of *Pelargonium odoratissimum* Ait.
     d) Extra jam, jam, extra jelly and jelly (in each made from apples,
        rosehips or quinces), sweetened chestnut purée: vanilla, vanilla
        extract, vanillin, ethyl vanillin.
  3) In addition to those ingredients specified in the reserved descriptions
     above, the following may contain the ingredients indicated.
     a) All products specified above: water (suitable for food
        manufacture), edible oils and fats, liquid pectin (derived from
        dried apple pomace and/or dried peel of citrus fruits, by the
        addition of dilute acid followed by partial neutralisation with
        sodium or potassium salts).
     b) Jam, reduced sugar jam: fruit juice.
     c) Extra jam, reduced sugar jam or reduced sugar marmalade (which
        in each case is made from rosehips, strawberries, raspberries,
        gooseberries, redcurrants and/or plums): red fruit juice.
     d) Jam, jelly, reduced sugar jam, reduced sugar jelly or reduced
        sugar marmalade (which in each case is made from strawberries,
        raspberries, gooseberries, redcurrants and/or plums): red beetroot
        juice.
     e) Jam, jelly, marmalade, reduced sugar products and UK standard
        jelly: permitted colours.
     f) Marmalade: essential oils of citrus fruits.
  4) Only the following sweetening agents may be used (except that

49

4.1

reduced sugar products and diabetic products (labelled as such) may contain sweeteners as permitted in the Sweeteners in Food Regulations):

dextrose anhydrous/monohydrate, dried glucose syrup, extra-white sugar, glucose syrup, invert sugar solution/syrup, semi-white sugar, sugar solution, white sugar, fructose, brown sugar, cane mollases, honey, aqueous solution of sucrose (meeting specified standards).

Soluble solids content (at 20°C)

	Soluble solids (%)
Extra jam, jam (1)	min. 60 (2)
Extra jelly, jelly (1)	min. 60 (2)
Marmalade (1)	min. 60 (2)
Sweetened chestnut purée (1)	min. 60 (2)
Reduced sugar jam, reduced sugar jelly, reduced sugar marmalade	30–55 (3)
UK standard jelly (1)	min. 60 (2)
X curd	min. 65
Y flavour curd	min. 65
Mincemeat	min. 65

Notes:
(1) Not applicable to 'diabetic' products labelled as such
(2) If the quantity is greater than a single serving, and the soluble solids is less than 63%, the pack must be marked 'keep in a cool place once opened'
(3) If the qunatity is greater than a single serving, the pack must be either marked 'keep in a cool place once opened' or contain sufficient preservative to have a preserving effect on the food

Labelling
1) 'Conserve' and 'preserve' may only be used for foods which are either 'jam' or 'extra jam'.
2) The word 'extra' may be dropped from 'extra jam' and 'extra jelly'.
3) Marmalade:
   a) with no insoluble matter (except a small quantity of finely sliced peel) may be called 'jelly marmalade',
   b) containing peel shall state the style of cut of the peel, and
   c) not containing peel shall state that it does not contain peel.
4) The name of the food shall, if a single type of fruit is used, include an indication of that fruit; if two types of fruit are used, include an indication of both fruits in descending order of weight; if more than two, either include an indication of all the fruits, or state 'mixed fruit', or state the number of fruits used.
5) All products must be marked with the following statements (except that products made specially for diabetics do not require the sugar statement):

'prepared with x g of fruit per 100 g'
'total sugar content: y g per 100 g'

with the appropriate figures substituted for x and y.

6   List of ingredients. The following points are specified:
    a) if three or more types of fruit are used, they can be listed together as 'fruit';
    b) if red beetroot juice is used as an ingredient, list shall state 'red beetroot juice to reinforce the colour';
    c) if any jam contains apricots which have been dried other than by freeze-drying, they shall be identified as 'dried apricots';
    d) where the residual sulphur dioxide content exceeds 30 mg/kg, the presence of sulphur dioxide shall be indicated in the list of ingredients.

General points

1) Any fruit used must:
   a) contain all its esential constituents and is sound, free from deterioration and sufficiently ripe for use; and
   b) have been cleaned and trimmed and have had any blemishes removed.
2) Any fruit juice must comply with the requirements for ingredients, processing and treatment specified in the Fruit Juices and Fruit Nectars Regulations
3) Fruit sources (other than fruit juice) may only be used if treated by being heated, chilled, frozen, freeze-dried or concentrated except:
   a) apricots may be used in jam if dried otherwise than by freeze-drying,
   b) ginger may have been dried or preserved in syrup, and
   c) citrus peel (used in extra jam, jam, marmalade, reduced sugar jam, reduced sugar marmalade, fruit curd, fruit flavour curd and mincemeat) may have been preserved in brine.
4) The term 'jelly' may be used for other products customarily known as jelly and which cannot be confused with products conforming to the specified standards.

Margarine

Regulation: Spreadable Fats (Marketing Standards) Regulations 1995 (1995/3116)

See details given under 'Spreadable fats'.

Meat

Specified bovine material

Regulation: Specified Bovine Materials (No. 2) Order 1996 (1996/1192)

Following the outbreak of bovine spongiform encephalopathy (BSE)

51

4.1

various controls have been introduced (and amended). This Regulation represents the main one at the time of writing which restricts the use of potentially harmful parts of infected animals. There are extensive controls now applied in this area and they are being regularly updated as new research evidence is available. Only a brief summary of this main control is therefore provided here.

Definition
Specified bovine material:

a) subject to (d) below, the head (including the brain but excluding the tongue), spinal cord, spleen, thymus, tonsils and intestines of a bovine animal 6 months old or over which has died in the UK or has been slaughtered there;
b) the thymus and intestines of a bovine animal 2 months old or over but less than six months old which has died in the UK or has been slaughtered there;
c) the thymus and intestines of a bovine animal under 2 months old which has been slaughtered in the UK for human consumption; and
d) subject to paragraph 4 of the regulation, the head (including the brain and the tongue) spinal cord, spleen, thymus, tonsils and intestines of a scheme animal, and includes anything left attached to such material after dissection of the carcase and any animal matter which comes into contact with the material after it has been removed from the carcase, but does not include the whole carcase.

Scheme animal: a bovine animal which has been slaughtered pursuant to the purchase scheme introduced under Commission Regulation 716/96.

Requirements

1) No person shall sell any specified bovine material or any food containing specified bovine material, for human consumption.
2) No person shall use any specified bovine material in the preparation of food for human consumption.
3) No person shall sell any specified bovine material for use in the preparation of food for human consumption.
4) No person shall use the vertebral column of a bovine animal in the recovery of meat by mechanical means.
5) No person shall use, in the preparation of food for sale for human consumption, any meat which has been recovered by mechanical means from the vertebral column of a bovine animal.
6) No person shall use the vertebral column of a bovine animal from which meat has been cut, to produce food other than fat or gelatin for human consumption.

General points
The Regulations contain controls relating to: registration requirements,

treatment, staining, handling of specified bovine material, rendering and the approval of collection centres, incinerators and rendering plants.

4.1

Meat products and spreadable fish products

Regulation: Meat Products and Spreadable Fish Products Regulations 1984 (1984/1566)
Amendments: Meat Products and Spreadable Fish Products (Amendment) Regulations 1986 (1986/987)
Sweeteners in Food Regulations 1995 (1995/3123)
Colours in Food Regulations 1995 (1995/3124)
Miscellaneous Food Additives Regulations 1995 (1995/3187)
Food Labelling Regulations 1996 (1996/1499)

Definitions

Meat: means the flesh, including fat, and the skin, rind, gristle and sinew in amounts naturally associated with the flesh used, of any animal or bird which is normally used for human consumption and includes the offals, obtained from such animal or bird, listed in part (a) of the definition of offals given below but does not include any other part of the carcase.

Lean meat: meat free, when raw, of visible fat.

Meat product: any food which consists of meat or has meat as an ingredient except:
- raw meat with no added ingredient except proteolytic enzymes;
- uncooked chickens, hens, cocks, turkeys, ducks, geese and guinea fowl (or their cuts or offals) with no added ingredient except additives permitted by the additive regulations, flavourings, smoke and smoke solutions, water, self-basting preparations or seasonings;
- haggis, black pudding, white pudding, brawn, collard head;
- sandwiches, filled rolls and similar bread products which are ready for consumption unless they contain meat and are sold using the words 'burger', 'economy burger' or 'hamburger';
- products named 'broth', 'gravy' or 'soup';
- stock cubes (and similar flavouring agents);
- products commonly known in Scotland as 'potted head', 'potted meat' and 'potted hough';
- products containing only animal or bird fat but no other meat.

Fish: the edible portion of any fish including edible molluscs and crustacea.

Spreadable fish product: any product containing fish in which 'paste', 'pâté' or 'spread' is used in the name or any other readily spreadable product containing fish (except if they only contain fish oil and no other fish constituent).

Offals (as used in the Food Labelling Regulations):   any of the following parts of the carcase:

4.1

a) for mammalian species – diaphragm, head meat (muscle meat and associated fatty tissue only), heart, kidney, liver, pancreas, tail meat, thymus, tongue;

b) for avian species – gizzard, heart, liver, neck;

c) for mammalian species – brains, feet, large intestine, small intestine, lungs, oesophagus, rectum, spinal cord, spleen, stomach, testicles, udder.

Cured meat: a food consisting of meat and curing salt and may include water, additives (other than flavourings, smoke and smoke solutions and subject to the additive regulations) and any of the following ingredients used in small quantities in accordance with good manufacturing practice but no other ingredient:

a) sucrose, invert sugar, glucose, dextrose, lactose, maltose and glucose syrup (all for sweetening purposes only);

b) honey, maple syrup, molasses, any food used solely as a garnish or decorative coating, flavourings, smoke and smoke solutions, and salt, herbs or spices used as seasoning;

c) hydrolysed proteins and yeast extracts (both for flavouring purposes only).

Curing salt: sodium chloride, potassium chloride, sodium nitrate, potassium nitrate, sodium nitrite or potassium nitrite whether alone or in any combination except that if only sodium chloride and potassium chloride are used (either alone or mixed) they must be used in a meat product in sufficient quantity to have a significant preserving effect.

Reserved descriptions

	Definition of X	Minimum meat content (% of food)	Minimum lean meat content (% of required meat content)	Minimum X content (% of food)
Burger (1):				
a) qualified by X	Named meat or	80	65	80
b) others	cured meat	80	65	
Economy burger (1):				
a) qualified by X	Named meat or	60	65	60
b) others	cured meat	60	65	
Hamburger (1, 2)		80	65	
Chopped X	'Meat', 'cured meat' or a named meat or cured meat	90	65	
Corned X	Named meat unless qualifed by the name of another food	120	96	

(*continued*)

	Definition of X	Minimum meat content (% of food)	Minimum lean meat content (% of required meat content)	Minimum X content (% of food)
Luncheon meat, luncheon X	Named meat or cured meat	80	65	
Meat pie, meat pudding, pie or pudding qualified by X, Melton Mowbray pie, game pie:	Named meat or cured meat unless qualified by the name of another food			
a) cooked weight				
>200 g		25	50	
>100 g, < 200 g		21	50	
<100 g		19	50	
b) uncooked weight				
>200 g		21	50	
>100 g, < 200 g		18	50	
<100 g		16	50	
Scottish pie or Scotch pie:				
a) cooked		20	50	
b) uncooked		17	50	
Pie or pudding qualified by X, pastie or pasty, bridie or sausage roll:	'Meat' or a named meat or cured meat and also the name of another food			
a) cooked		12.5 (4)	50	
b) uncooked		10.5 (4)	50	
Sausage, chipolata, link or sausage meat:				
a) qualified by X	'Pork' and no other meat	65	50	80 (5)
	'Beef' and no other meat	50	50	50 (5)
	'Liver' and/or 'tongue'	50	50	30 (5)
	named meat or cured meat	50	50	80 (5)
b) unqualified		50	50	
Paste or pâté (unless preceded by words which include the name of a food other than meat or fish and do not include a named meat, fish, cured meat or fish or 'meat' or 'fish':				
a) if preceded by X	named meat, fish cured meat			(6)

and if
(*continued*)

	Definition of X	*Minimum meat content (% of food)*	*Minimum lean meat content (% of required meat content)*	*Minimum X content (% of food)*
b) a meat product				
– pâté		70	50	
– others		70	65	
c) a spreadable fish product		70 (7)		
d) combined meat and spreadable fish product				
– pâté		70 (8)	50	
– others		70 (8)	65	
Spread, with the same condition as for paste except that the name must be preceded by X and if a meat product	Named meat, fish or cured meat or fish or a mixture of them	70	65	70

Notes:
   (1) Whether or not the description forms part of another word except that for 'burger' excluding names using 'economy burger' or 'hamburger'. Where the name refers to a compound product (e.g. including a roll) the % only applies to the meat component of the product
   (2) Meat must be beef and/or pork and the name of the food must be qualified by the type of meat used
   (3) The food shall only contain corned meat of the named type
   (4) If the name used is pie or pudding qualified by 'meat' or a named meat or cured meat and preceded by 'vegetable' or the name of a type of vegetable, the minimum meat content is 10% (cooked) or 8% (uncooked)
   (5) As % of meat content
   (6) Sufficient so that the food is characterised by X
   (7) Fish content
   (8) Sum of fish and meat contents

Specific labelling requirements
   1) Every meat product (except those to which (2) and (3) apply but not those requiring an indication of other ingredients under (3) (a)) or spreadable fish product must be labelled with a declaration 'minimum x% meat' or 'minimum x% fish' with the correct value being inserted for x. Where a product includes a liquid medium not normally consumed the declaration must indicate that the liquid weight is not included in the calculation. If x is less than 10% the declaration may be 'less than 10% meat' or 'less than 10% fish'. If x is more than 100% the declaration shall be 'not less than 100% meat' or 'not less than 100% fish'. The words meat or fish may be replaced by the name of the meat or fish used (see also (4), (5) and (7)).

2) For corned meat products, if only corned meat is used the declaration shall be '100% corned X' where the type of meat is inserted for X. Where the product name is qualified by the name of another food or where 'corned X' is used in the name of a food, the declaration should read 'minimum y% corned X' with the correct value being inserted for y (see also (4)).

3) For any product having the appearance (while disregarding the presence of any seasoning, flavouring, garnishing or gelatinous substance) of a cut, joint, slice, portion or carcase of raw meat, cooked meat or cured meat the following shall apply (except in the case of products using the reserved descriptions listed above or food appearing to be shaped minced raw meat):

   a) The name used shall include an indication of any ingredient other than water and other than
      i) for cooked meat – additives (permitted by the additive regulations, flavourings, smoke and smoke solutions), salt and herbs or spices used as seasoning;
      ii) for cured meat – any ingredient listed in the definition of cured meat given above and used according to the conditions specified.

   b) If under (a) for raw or cooked meat containing added water or uncooked cured meat containing more than 10% added water the name includes an indication of other ingredients, then it shall be labelled with the declaration 'with added water'.

   c) If under (a) the name does not include an indication of other ingredients but it does contain added water (or for uncooked cured meats, contains more than 10% added water) it should be labelled with a declaration 'with not more than x% added water' where in the case of:
      – raw or cooked meat, x is the value of the maximum added water;
      – uncooked cured meat, x is the value expressed as a multiple of 5 by which the maximum added water exceeds 10%;
      – cooked cured meat, x is the value expressed as a multiple of 5 of the maximum added water.

   Declarations under (b) and (c) shall appear either in the name of the food or in immediate proximity to it.

4) Declarations under (1) and (2) shall, if the product is labelled with an ingredients list, appear in immediate proximity to that list. Where only one type of meat or cured meat, fish or cured fish is present it may appear in immediate proximity to the relevant ingredient in the ingredients list and (except for cured meat) the word 'meat' or 'fish' or the name of the meat or fish may be omitted from the declaration.

5) Products requiring a declaration under (1) must have a lean meat content of at least 65% of the declared minimum meat content except the following products which must have at least 50%: meat pie or

4.1

part of a meat pie (includes meat pudding or sausage roll), a sausage (including chipolata, frankfurter, link, salami and any similar products and includes sausage meat) or a meat product of which sausage is an ingredient, or pâté.

6) Products requiring a declaration under (2) must have a lean meat content of at least 96% of the meat content of the product.

7) Where a meat product has a declaration under (1) or (2) and has animal or bird fat which, because of (5) or (6) cannot be taken into account in that declaration then, subject to (8), that fat must be separately identified in any ingredients list.

8) If the declared meat content is below that actually present and thus the fat cannot be taken into account for the declaration then the fat can be disregarded for the purposes of (7) if the lean meat contents specified in (5) are met.

General points

1) Meat content: the meat content is the sum of meat (calculated as raw meat) used in the preparation of the product and, if its presence is indicated in the name of the food, any naturally associated solid bone expressed as a % of the total weight of the product as sold. Where the product is dehydrated, or contains dehydrated ingredients, and is reconstituted before consumption, the calculation shall use the reconstituted weights. For sausages, any edible skin enclosing the sausage shall not be included in either the weight of the meat used or the product weight (where sausage includes chipolata, frankfurter, link, salami and any similar product and sausage meat). For products which include a liquid medium not normally consumed the weight of liquid shall not be included in the weight of the product.

2) Fish content: the fish content is the total weight of fish (calculated as raw fish) used in the preparation of the product expressed as a % of the total weight of the product as sold.

3) Corned meat content: the corned meat content shall be taken to be five-sixths of the meat content of the product.

4) Added water content: the added water content is the total weight of added water in the product expressed as a % of the total weight of the product as sold. The added water is only that water which is added in excess of that which would naturally be present in the meat used in the product when raw.

5) Use of offals: only those offals listed in (a) in the definition of offals given above may be used in the preparation of uncooked meat products unless used solely as a sausage skin (which includes chipolata, frankfurter, link, salami and any similar product and sausage meat). (See also separate section on 'Use of bovine offals'.)

6) The Regulations do not apply to foods marked or labelled as being intended exclusively for consumption by babies or young children.

Milk

A Fresh liquid milk

4.1

Regulations: EEC Council Regulation 1411/71
  Drinking Milk Regulations 1976 (1976/1883)
  Amendments: EEC Council Regulations 1556/74, 556/76, 222/88, 2138/92
    Milk and Dairies (Standardisation and Importation) Regulations 1992 (1992/3143)
    Dairy Products (Hygiene) Regulations 1995 (1995/1086)

Definitions
Milk: the milk yield of one or more cows.
Drinking milk: the following products for delivery to the consumer –
  raw milk, whole milk, semi-skimmed milk and skimmed milk.
Raw milk: milk which has not been heated or subjected to treatment
  having the same effect.
Whole milk: milk which has been subjected to at least one heat treatment
  or an authorised treatment of equivalent effect by a milk processor, and
  with respect to fat content, meets the requirements of the table below.
Semi-skimmed milk/skimmed milk: milk which has been subjected to at
  least one heat treatment or an authorised treatment of equivalent effect
  by a milk processor, and the fat content of which meets the requirements
  of the table below.

	Fat content (%)
Whole milk:	
non-standardised	3.5 (1, 2)
standardised	3.5
Semi-skimmed milk	1.5–1.8
Skimmed milk	max. 0.3

Notes:
  (1) Fat content must not have been altered since the milking stage either by the
      addition or separation of milk fat or by mixture with milk, the natural fat
      content of which has been altered
  (2) The EEC Regulation allows for the granting of a derogation for up to 1 year
      for areas where the milk produced does not meet the required fat content.
      However it must be min. 3.2%

Requirements
No person shall sell drinking milk except:
  a) raw milk,
  b) non-standardised whole milk,
  c) standardised whole milk,
  d) semi-skimmed milk,
  e) skimmed milk.

General points
No person shall make any alteration to the composition of drinking milk.

59

4.1

Except that standardised whole milk, skimmed milk or semi-skimmed milk may be obtained by adding or separating milk or cream, or by adding skimmed or semi-skimmed milk.

B   Condensed and dried milk

Regulation: Condensed Milk and Dried Milk Regulations 1977 (1977/928)
  Amendments: Food Labelling Regulations 1980 (1980/1849)
  Condensed Milk and Dried Milk (Amendment) Regulations 1982 (1982/1066)
  Condensed Milk and Dried Milk (Amendment) Regulations 1986 (1986/2299)
  Condensed Milk and Dried Milk (Amendment) Regulations 1989 (1989/1959)
  Miscellaneous Food Additives Regulations 1995 (1995/3187)
  Food Labelling Regulations 1996 (1996/1499)

Definitions
Condensed milk: milk, partly skimmed milk or skimmed milk (or any combination) whether with or without the addition of cream, dried milk or sucrose, which has been concentrated by the partial removal of water but does not include dried milk.
Dried milk: milk, partly skimmed milk or skimmed milk (or any combination) whether with or without the addition of cream, which has been concentrated to the form of powder, granule or solid by the removal of water.

Reserved descriptions

	Milk fat (%)	Minimum total milk solids (%) (1)	Maximum moisture (%)
Unsweetened condensed high-fat mik (2)	min. 15	26.5	
Evaporated mik (2)	9–15	31	
Unsweetened condensed mik (2)	7.5–15	25	
Unsweetened condensed partly skimmed milk (2):			
a) retail sale	4–4.5	24	
b) other	1.0–7.5	20	
Unsweetened condensed skimmed milk (2)	max. 1.0	20	
Sweetened condensed milk (3):			
a) retail sale	min. 9	31	
b) other	min. 8	28	
Sweetened condensed partly skimmed milk (3):			
a) retail sale	4–4.5	28	
b) other	1.0–8	24	
Sweetened condensed skimmed milk (3)	max. 1.0	24	
Dried high-fat milk, high-fat milk powder	42–65		5
Dried whole milk, whole-milk powder	26 –42		5
Dried partly skimmed milk, partly skimmed-milk powder	1.5–26		5
Dried skimmed milk, skimmed-milk powder	max. 1.5		5

See next page for Notes

4.1

Notes:
  (1) For all products, up to 0.3% of milk solids not fat can consist of lactose; for condensed products, up to 25% of total milk solids can be derived from dried milk
  (2) Condensed milk which has been sterilised
  (3) Condensed milk with added sucrose

Permitted ingredients
  1) Any condensed-milk product or dried-milk product may contain any miscellaneous additive permitted by the Miscellaneous Food Additives Regulations.
  2) Sweetened condensed-milk products may contain up to 0.02% added lactose.
  3) Condensed- and dried-milk products may contain added vitamins.

General points
  1) Any milk source used or any final product must be subject to pasteurisation so that a specified phosphatase test is satisfied.
  2) The following declarations are required for retail sale:
    a) for condensed milk (minimum 1.0% milk fat) and dried milk (minimum 1.5% milk fat) – % actual or minimum milk fat content;
    b) any instant dried-milk product containing added lecithins (E322) – 'instant';
    c) for condensed milk – % actual or minimum milk solids not fat;
    d) for condensed milk – instructions for dilution or reconstitution or use (when intended for use in unaltered state);
    e) for dried milk (minimum 1.5% milk fat) – fat content after dilution or reconstitution and instructions for dilution or reconstitution;
    f) for unsweetened condensed UHT products – 'UHT' or 'ultra-heat-treated';
    g) for condensed milk (maximum 1.0% milk fat) and dried milk (maximum 1.5% milk fat) – the statement 'not to be used for babies except under medical advice'.
  3) Condensed or dried milk specially prepared for infant feeding and labelled as such is exempt from the above controls on additional ingredients and labelling.

Olive oil

Regulations: Olive Oil (Marketing Standards) Regulations 1987 (1987/1783)
  Tetrachloroethylene in Olive Oil Regulations 1989 (1989/910)
Amendment: Olive Oil (Marketing Standards)(Amendment) Regulations 1992 (1992/2590)

Note: The Regulations provide for procedures to enforce marketing standards applied by EC Council Regulation Nos 136/66 and Commission Regulation 1860/88 as amended by Nos 1915/87 and 356/92. The

4.1 standards have been supplemented by very detailed controls on characteristics and methods of analysis contained in Commission Regulation 2568/91. The use of the following definitions (and associated standards) is compulsory in the marketing of olive oil and olive–pomace oil within the UK, in intra-Community trade and in trade with third countries.

Definitions
Note: All the following definitions are supplemented by the statement 'and the other characteristics which comply with those laid down for this category'. Reference should be made to Commission Regulation 2568/91 for the more detailed characteristics.

Virgin olive oils: oils derived solely from olives using mechanical or other physical means under conditions, and particularly thermal conditions, that do not lead to deterioration of the oil, and which have undergone no treatment other than washing, decantation, centrifugation and filtration, but excluding oils obtained by means of solvents or of re-esterification and mixtures with other oils.

Refined olive oil: olive oil obtained by refining virgin olive oils.

Olive oil: olive oil obtained by blending refined olive oil and virgin olive oil other than lampante oil.

Crude olive-residue oil: oil obtained by treating olive residue with solvents, but excluding oil obtained by re-esterification and mixtures with other types of oils.

Refined olive-residue oil: oil obtained by refining crude olive-residue oil.

Olive-residue oil: oil obtained by blending refined olive-residue oil and virgin olive oil other than lampante oil.

The following standards are included with the definitions:

	Organoleptic grading (1)	*Free acid content (expressed as oleic acid)*(g/100 g)
Virgin olive oils:		
extra virgin olive oil	min. 6.5	max. 1
virgin olive oil (2)	min. 5.5	max. 2
ordinary virgin olive oil	min. 3.5	max. 3.3
lampante virgin olive oil	less than 3.5	more than 3.3
Refined olive oil		max. 0.5
Olive oil		max. 1.5
Refined olive-residue oil		max. 0.5
Olive-residue oil		max. 1.5

Notes:
(1) As measured according to the procedure specified in Commission Regulation 2568/91

(2) The expression 'fine' may be used at the production and wholesale stage

(3) Only one of the specified levels needs to be met for the oil to be in this category

4.1

General points

1) Only extra-virgin olive oil, virgin olive oil, olive oil and olive-residue oil may be sold by retail.

2) No olive oil or olive-residue oil with a tetrachloroethylene content of more than 0.1 mg/kg can be sold by retail.

Poultry (frozen, water content)

Regulations: Poultry Meat (Water Content) Regulations 1984 (1984/1145)
EEC Council/Commission Regulations 2967/76, 2785/80
Amendments: EEC Council/Commission Regulations 1691/77, 641/79, 2632/80, 2835/80, 3204/83.

Frozen and deep-frozen chickens, hens and cocks may be marketed only if the water content absorbed during preparation does not exceed the technically unavoidable minimum determined by specified methods.

Products treated with polyphosphates are exempt provided they are clearly described with their treatment.

Products may be described as 'dry chilled poultry' if no water is added during chilling.

	Maximum extraneous water (%)
Slaughterhouse check	
prior to freezing	5
Rapid detection check (thawing loss) (1):	
standard freezing	5.2 (2)
dry chill poultry	2 (2)
Standard procedure (analytical) (3):	
standard freezing	(4)
dry chill poultry	(4)

Notes:

(1) Based on a sample of 20 carcases. Not to be used if polyphosphates have been used to increase water retention

(2) If these figures are exceeded, then the standard procedure should be followed.

(3) Based on a sample of seven carcases

(4) Results obtained by comparing two figures obtained by calculation (see amended Annex III of Council Regulation 2967/76)

Spreadable fats

Regulation: Spreadable Fats (Marketing Standards) Regulations 1995 (1995/3116)
EC Council Regulation 2991/94

Only certain products meeting certain specified standards may be marketed under the terms specified by these controls. The controls relate

4.1 to milk fats, fats, and fats composed of plant and/or animal products falling within certain specified reference codes (see EC Regulation).

The controls apply to products
a) with a fat content of 10–90%,
b) with the fat content excluding salt being at least two-thirds of the dry matter,
c) which remain solid at 20°C, and
d) which are suitable for use as spreads.

Definitions

Milk fats: products in the form of a solid, malleable, emulsion. principally of the oil-in-water type, derived exclusively from milk and/or certain milk products, for which the fat is the essential constituent of value. However, other substances necessary for their manufacture may be added, provided those substances are not used for the purpose of replacing, either in whole or in part any milk constituent.

Fats: products in the form of a solid, malleable, emulsion, principally of the oil-in-water type, derived from solid and/or liquid vegetable and/or animal fats suitable for human consumption, with a milk fat content of not more than 3% of the fat content.

Fats composed of plant and/or animal products: products in the form of a solid, malleable, emulsion, principally of the oil-in-water type, derived from solid and/or liquid vegetable and/or animal fats suitable for human consumption, with a milk fat content of between 10 and 80% of the fat content.

Vitamin A: vitamin A present as such or as its esters and includes β-carotene on the basis that 6 μg of β-carotene or 12 μg of other biologically active carotenoids equal 1 μg of retinol equivalent.

Vitamin D: the anti-rachitic vitamins.

Reserved descriptions

The following descriptions are reserved for products meeting the specification. However, the requirements do not apply to:
a) products the exact nature of which is clear from traditional usage and/or when the designations are clearly used to describe a characteristic quality of the product; or
b) concentrated products (butter, margarine, blends) with a fat content of 90% or more.

Fat group and reserved description	Additional specified requirements			4.1
	Milk fat content (%)	Water content (%)	Dry non-fat milk-material content (%)	
Milk fats:				
Butter	80–90	max. 16%	max. 2%	
Three-quarter-fat butter	60–62 (1)			
Half-fat butter	39–41 (2)			
Dairy spread x%	a) less than 39 (2) b) 41–60 (1) c) 62–80			
Fats:				
Margarine (3)	80–90			
Three-quarter-fat margarine (3)	60–62 (1)			
Half-fat margarine or minarine or halvarine (3)	39–41 (2)			
Fat spreads x% (3)	a) less than 39 (2) b) 41–60 (1) c) 62–80			
Fats composed of plant and/or animal products:				
Blend	80–90			
Three-quarter-fat blend	60–62 (1)			
Half-fat blend	39–41 (2)			
Blended spread x%	a) less than 39 (2) b) 41–60 (1) c) 62–80			

Notes:

(1) The description of these products may be accompanied by the term 'reduced-fat'. If used for a 'three-quarter-fat' product , it may replace the term 'three-quarter-fat'

(2) The description of these products may be accompanied by the term 'low-fat' or 'light'. If used for a 'half-fat' product , it may replace the term 'half-fat'

(3) The term 'vegetable' may be used provided that the product contains only fat of vegetable origin with a tolerance of 2% of the fat content for animal fats. If reference is made to a vegetable species the same tolerance applies

Vitamin content of margarine

Any margarine sold by retail shall contain the following:

a) vitamin A, 800–1000 µg per 100 g margarine;

b) vitamin D, 7.05–8.82 µg per 100 g margarine.

This requirement does not apply to margarine brought into Great Britain from an EEA State in which it had been lawfully produced and sold or from an EC Member State in which it was in free circulation and lawfully sold. The margarine must be suitably labelled to give the nature of the margarine.

4.1 *Labelling requirements*

1) In addition to the general labelling requirments for foods, the following are required for products using the above reserved descriptions:
   a) the reserved description;
   b) the total % fat content by weight at the time of production;
   c) for fats composed of plant and/or animal products, the vegetable, milk or other animal fat content in decreasing order of weighted importance (as % total weight at time of production); and
   d) the % salt content (indicated in a particularly legible manner).
2) The reserved description may be accompanied by terms:
   a) to define the plant and/or animal species from which the products originate;
   b) to define the intended use of the product; or
   c) to indicate the production methods.
3) The term 'traditional' may be used with the description 'butter' where the product is obtained directly from milk or cream (where, in this case, 'cream' is defined as: the product obtained from milk in the form of an emulsion of the oil-in-water type with a milk-fat content of at least 10%).
4) Only those terms referred to in notes (1) and (2) of the preceding table may be used to state, imply or suggest fat content of the products in the table. However, if a term was legally being used on 31 December 1993 it can continue to be used until 9 December 1999.

General points

The Regulations do not apply to any spreadable fat brought into Great Britain from an EEA State (other than an EC Member State) in which it had been lawfully produced and sold. The spreadable fat must be suitably labelled to give the nature of the spreadable fat.

Sugar products

Regulation: Specified Sugar Products Regulations 1976 (1976/509)
Amendments: Miscellaneous Additives in Food Regulations 1980 (1980/1834)
  Food Labelling Regulations 1980 (1980/1849)
  Specified Sugar Products (Amendment) Regulations 1982 (1982/255)
  Colours in Food Regulations 1995 (1995/3124)
  Miscellaneous Food Additives Regulations 1995 (1995/3187)
  Food Labelling Regulations 1996 (1996/1499)

Definitions

Candy sugar: crystalline sugar with crystals having any dimension greater than 1 cm.

Loaf sugar: a piece of agglomerated crystalline sugar, usually conically shaped, weighing not less than 250 g.

66

Reserved descriptions
(Descriptions to be applied only to the products stated and conforming to the standards below.)

4.1

Dextrose anhydrous: purified and crystallised D-glucose (except icing dextrose) containing no water of crystallisation.

Dextrose monohydrate: purified and crystallised D-glucose (except icing dextrose) containing one molecule of water of crystallisation.

Glucose syrup: purified and concentrated aqueous solution of nutritive saccharides obtained from starch.

Dried glucose syrup: glucose syrup which has been partially dried.

Extra-white sugar/white sugar/sugar: purified and crystallised sucrose (except icing sugar, candy sugar or loaf sugar).

Semi-white sugar: purified and crystallised sucrose (except candy sugar or loaf sugar).

Sugar solution: an aqueous solution of sucrose.

Invert sugar solution: an aqueous solution of sucrose which has been partially inverted by hydrolysis.

Invert sugar syrup: an aqueous solution, whether or not crystallised, of sucrose which has been partially inverted by hydrolysis.

White soft sugar/soft sugar: fine-grain purified moist sucrose.

Lactose: the carbohydrate normally obtained from whey; it may be anhydrous or contain one molecule of water of crystallisation or be a mixture of both forms.

Icing dextrose/powdered dextrose: fine particles of dextrose monohydrate or anhydrous (or mixture).

Icing sugar/powdered sugar: fine particles of white sugar or extra-white sugar (or mixtures).

	Minimum dry matter (%)	*Minimum D-glucose content (% of dry matter)*	*Maximum sulphated ash content (% of dry matter)*
Dextrose anhydrous	98	99.5	0.25
Dextrose monohydrate	90	99.5	0.25
Dried glucose syrup	93	20 (1)	1.0
Glucose syrup	70	20 (1)	1.0

	Minimum dry matter (%)	*Invert sugar content (% of dry matter)*	*Invert sugar fructose/ dextrose ratio*	*Maximum conductivity ash content (% of dry matter)*
Sugar solution (2,5)	62	max. 3	1.0 ± 0.2	0.1
Invert sugar solution (5)	62	3–50	1.0 ± 0.1	0.4
Invert sugar syrup (5)	62	min. 50	1.0 ± 0.1	0.4

4.1

	Minimum sucrose and invert sugar content as sucrose (%)	Invert sugar content (%)	Maximum ash content (%)	Maximum loss on drying (%)
White soft sugar (2)	97	0.3–12	0.2 (3)	3
Soft sugar	88	0.3–12	3.5 (4)	4.5

	Minimum polarization (°)	Maximum invert sugar (%)	Maximum loss on drying (%)
Extra-white sugar (2)	99.7	0.04	0.1
White sugar, sugar (2)	99.7	0.04	0.1
Semi-white sugar	99.5	0.1	0.1

	Minimum anhydrous lactose (% dry matter)	Maximum ash content (%)	Maximum loss on drying (%)	pH (10% solution at 20°C)
Lactose	97	0.8 (4)	6	4.5–7.0

Notes:
(1) Minimum dextrose equivalent expressed as D-glucose (% of dry matter)
(2) Extra standards are specified for colour standards (see Regulations)
(3) Conductivity ash content
(4) Sulphated ash content
(5) May only be qualified by the word 'white' if certain colour standards are satisfied (see Regulations)

Permitted additional ingredients
1) Any specified sugar product may contain miscellaneous additives (subject to the Miscellaneous Food Additives Regulations 1995).
2) Icing sugar or icing dextrose may contain maximum 5% starch if it does not contain a miscellaneous additive used primarily as an anti-caking agent.
3) If a sugar product is intended for use as an ingredient of another food it may contain colour (subject to the Colours in Food Regulations 1995).

Labelling requirements
1) Extra-white sugar may be called 'sugar' or 'white sugar'.
2) Dextrose monohydrate or dextrose anhydrous may be called 'dextrose' when sold by retail (or consigned or delivered for retail sale).
3) Any invert sugar syrup containing a significant proportion of crystals in solution must be qualified by the word 'crystallised' (or similar).

4) For sugar solution, invert sugar solution or syrup, a declaration of the dry matter and of the invert sugar content must be given.

4.1

5) For glucose syrup or dried glucose syrup containing more than 20 mg/kg sulphur dioxide, a declaration that the product is not for retail sale (and a declaration on an accompanying document of the proportion of sulphur dioxide present).

6) Any sugar product containing permitted added colour (see above), a declaration 'contains permitted colour' or 'contains colour' or 'contains X' where X is the designation of the colour.

7) For icing sugar or icing dextrose containing either any starch or any miscellaneous additive used as an anti-caking agent, a declaration 'contains starch' or 'contains X' where X is the designation of the anti-caking agent.

8) For glucose syrup or dried glucose syrup containing any miscellaneous additive used as an anti-foaming agent, the declaration 'contains X' where X is the designation of the anti-foaming agent.

Water

A Drinking water

Regulation: Drinking Water in Containers Regulations 1994 (1994/743)

These Regulations do not apply to a recognised natural mineral water or a medicinal product (as provided by the Medicines Act 1968).

Definitions

Drinking water: water intended for sale for human consumption in any bottle.

Bottle: (as a noun) a closed container of any kind in which water is sold for human consumption; (as a verb) construed accordingly.

Requirements

Detailed requirements are specified for a range of properties, elements, organisms and substances and are given in the table overleaf. In addition, the Regulations specify the following:

1) The drinking water shall not contain any property, element, organism or substance
   a) not listed below – at a concentration or value which would be injurious to health,
   b) whether or not listed below – at a concentration or value which in conjunction with any other property, element, organism or substance it contains would be injurious to health.

2) Maximum concentration of trihalomethanes: $100\,\mu g/l$.

3) If water softened or desalinated: minimum hardness 60 mg Ca/l, minimum alkalinity 30 mg HCO_3/l.

4.1

	Unit of measurement	Maximum value (unless otherwise specified)
A Colour	mg/l Pt/Co scale	20
Turbidity (including suspended solids)	Formazin turbidity units	4
Odour (including hydrogen sulphide)	Dilution number	3 at 25°C
Taste	Dilution number	3 at 25°C
Temperature	°C	25
Sulphate	mg SO_4/l	250
Magnesium	mg Mg/l	50
Sodium	mg Na/l	150
Potassium	mg K/l	12
Dry residues	mg/l	1500 (after drying at 180°C)
Nitrate	mg NO_3/l	50
Nitrite	mg NO_2/l	0.1
Ammonium (ammonia and ammonium ions)	mg NH_4/l	0.5
Kjeldahl nitrogen	mg N/l	1
Oxidisability (permanganate value)	mg O_2/l	5
Total organic carbon	mg C/l	No significant increase over that normally observed
Dissolved or emulsified hydrocarbons (after extraction with petroleum ether; mineral oils	μg/l	10
Phenols	μg C_6H_5OH/l	0.5
Surfactants	μg/l (as lauryl sulphate)	200
Aluminium	μg Al/l	200
Iron	μg Fe/l	200
Manganese	μg Mn/l	50
Copper	μg Cu/l	3000
Zinc	μg Zn/l	5000
Phosphorus	μg P/l	2200
Fluoride	μg F/l	1500
Silver	μg Ag/l	10 (or, if silver used as water treatment agent: 80)
B Arsenic	μg/l	50
Cadmium	μg/l	5
Cyanide	μg/l	50
Chromium	μg/l	50
Mercury	μg/l	1
Nickel	μg/l	50
Lead	μg/l	50
Antimony	μg/l	10
Selenium	μg/l	10

(*continued*)

	Unit of measurement	Maximum value (unless otherwise specified)
Pesticides and related products (1):		
(a) individual substances	µg/1	0.1
(b) total substances (sum of total detected concentrations of individual substances)	µg/1	0.5
Polycyclic aromatic hydrocarbons (2)	µg/1	0.2
C Total coliforms	Number/100 ml	0
Faecal coliforms	Number/100 ml	0
Faecal streptococci	Number/100 ml	0
Sulphite-reducing clostridia	Number/20 ml	≤ 1
Colony counts	Number/1 ml at 22°C	100 (3)
	Number/1 ml at 37°C	20 (4)
D Conductivity	µS/cm	1500 at 20°C
Chloride	mg Cl/1	400
Calcium	mg Ca/1	250
Substances extractable in chloroform	mg/1 dry residue	1
Boron	µg B/1	2000
Barium	µg Ba/1	1000
Benzo[3, 4]pyrene	µg/1	10
Tetrachloromethane	µg/1	3
Trichloroethene	µg/1	30
Tetrachloroethene	µg/1	10

Notes:
(1) Pesticide means any fungicide, herbicide or insecticide and polychlorinated biphenyls and terphenyls
(2) The sum of the detected concentrations of fluoroanthene, benzo[3,4]fluoroanthene, benzo[11,12] flouroanthene, benzo[3,4]pyrene, benzo[1,12]perylene and indeno[1,2,3-cd]pyrene
(3) Analysis by multiple tube method
(4) The total viable colony count should be measured within 12 hours of bottling with the sample water being kept at a constant temperature during that 12-hour period. Any increase in the total viable count of the water between 12 hours after bottling and the time of sale shall not be greater than that normally expected

B Natural mineral waters

Regulation: Natural Mineral Waters Regulations 1985 (1985/71)
Amendment: Food Labelling Regulations 1996 (1996/1499)

Definitions
Natural mineral water: water which originates in a groundwater body or deposit and is extracted for human consumption from the ground through a spring, well, bore or other exit, and which is recognised in accordance with the Regulations.

4.1

Effervescent natural mineral water: a natural mineral water which spontaneously and visibly gives off carbon dioxide under ambient conditions of temperature and pressure.

Carbonated natural mineral water: an effervescent natural mineral water whose carbon dioxide content derives at least in part from an origin other than the groundwater body or deposit from which the water comes.

Naturally carbonated natural mineral water: an effervescent natural mineral water whose carbon dioxide content is the same after decanting (if it is decanted) and bottling as it was at source, and includes a natural mineral water to which carbon dioxide from the same ground body or deposit as the water has been added if the amount added does not exceed the amount previously released during decanting or bottling.

General points

1) To be classed as a natural mineral water, the water must originate in a groundwater body or deposit and be extracted for human consumption from the ground through a spring, well, bore or other exit and be officially recognised. Official recognition can be obtained by demonstrating that the list of conditions specified in the Regulations are satisfied and by providing relevant details of the source. These include: hydrogeological description of the source, physical and chemical characteristics of the water, microbiological analyses, levels of toxic substances, freedom from pollution and stability of the source. The names of all natural mineral waters which have been officially recognised will be published in the *Official Journal of the European Communities*.

2) Certain general requirements apply to the exploitation of a source of natural mineral water. These Regulations specify that equipment must be installed so as to avoid any possibility of contamination and to preserve the properties ascribed to it which the water possesses at source. In particular:

   a) the source must be protected against risks of pollution;

   b) the equipment for water extraction, pipes and reservoirs shall be of materials suitable for their purpose, and so made as to minimise chemical, physico-chemical or microbiological alteration of the water;

   c) containers to be so treated or manufactured as to minimise effects on the microbiological and chemical characteristics of the water;

   d) the washing and bottling plant and all other aspects of exploitation shall be of a satisfactory hygienic standard;

   e) the water must be transported from the source in the containers in which it is to be sold to consumers (unless prior to 17 July 1980 it was being transported in tanks to a bottling plant in the UK away from the source).

4.1

3) Authorised treatments: no treatment is permitted whether for disinfection or for any other purpose, except:
   a) filtration or decanting, preceded if necessary by oxygenation, provided that it does not alter the composition in respect of the stable constituents and it is not intended to change the total viable colony count;
   b) the total or partial elimination of carbon dioxide by exclusively physical methods;
   c) the addition of carbon dioxide.

4) Microbiological requirements: the following microbiological criteria must be satisfied:
   a) At source – total viable colony count to conform to normal values indicating the source protected from contamination.
   b) After bottling (within 12 hours) – the total viable colony count to be: 100 per ml at 20–22°C in 72 hours, and 20 per ml at 37°C in 24 hours; thereafter, no more than that which results from the normal increase.
   c) At source and thereafter – water to be free from:
      – parasites and pathogenic micro-organisms,
      – *Escherichia coli* and other coliforms and faecal steptococci (in any 250-ml sample),
      – sporulated sulphite-reducing anaerobes (in any 50-ml sample),
      – *Pseudomonas aeruginosa* (in any 250-ml sample).

5) Toxic substances: the following maximum are specified and the water must not contain any other substance in amounts which makes it unwholesome: mercury 1 g/l; cadmium 5 g/l; antimony, selenium, lead 10 g/l; arsenic, cyanide, chromium, nickel 50 g/l.

6) Organoleptic defects: no natural mineral water which has any organoleptic defect may be bottled or sold.

7) Bottles: the water must be sold in bottles which:
   a) are the ones originally used at the time of bottling by the exploiter,
   b) were fitted at the time of filling with closures designed to avoid any possibility of adulteration or contamination, and
   c) are still fitted with the intact original closures.

Specific labelling requirements
1) Every natural mineral water shall be labelled with:
   a) if non-effervescent, 'natural mineral water';
   b) if effervescent, 'naturally carbonated natural mineral water', 'natural mineral water fortified with gas from the spring' or

4.1

'carbonated natural mineral water' as appropriate (see definitions above);

c) if the water has been treated to remove carbon dioxide, 'fully de-carbonated' or 'partially de-carbonated' as appropriate;

d) the name of the place where the source is exploited and the name of the spring, well or borehole;

e) either the statement 'composition in accordance with the results of the officially recognised analysis of ...' (giving the date of the analysis), or a statement of the analytical composition of the water including details of its characteristic constituents.

2) A commercial designation (name under which water is sold including brand name, trade mark or fancy name) may be used but:

a) the name of a locality, hamlet or other place may only be used if it is where the source is exploited and its use does not mislead;

b) only one commercial designation may be used per source (except the brand name, trade mark or fancy name may differ);

c) the name of the source must be at least one and a half times the height and width of the largest letters used for any other part of the commercial designation.

3) No bottled drinking water may be labelled in a way likely to be confused with a natural mineral water.

4) No labelling must suggest a characteristic which the water does not possess. The following indications may only be used if they meet the conditions specified and in particular no other indication relating to the prevention, treatment or cure of human disease shall be used:

Indication	Constituent	Requirement
'Low mineral content'	Inorganic constituents	Not above 500 mg dry residue/l
'Very low mineral content'	Inorganic constituents	Not above 50 mg dry residue/l
'Rich in mineral salts'	Inorganic constituents	Above 1500 mg dry residue/l
'Contains bicarbonate'	Bicarbonate	Above 600 mg/l
'Contains sulphate'	Sulphate	Above 200 mg/l
'Contains chloride'	Chloride	Above 200 mg/l
'Contains calcium'	Calcium	Above 150 mg/l
'Contains magnesium'	magnesium	Above 50 g/l
'Contains fluoride'	Fluoride	Above 1 mg/l
'Contains iron'	Bivalent iron	Above 1 mg/l
'Acidic'	Free carbon dioxide	Above 250 mg/l
'Contains sodium'	Sodium	Above 200 mg/l
'Suitable for a low sodium diet'	Sodium	Not above 20 mg/l

4.2 Additives

Colours

Regulation: Colours in Food Regulations 1995 (1995/3124)

Definitions

Colour (as a noun):   a food additive which is used or intended to be used for the primary purpose of adding or restoring colouring in a food, and includes
  a) any natural constituent of food and any natural source not normally consumed as food as such and not normally used as a characteristic ingredient of food, and
  b) any preparation obtained from food or from any other natural source material, by, in either case (i) physical extraction, (ii) chemical extraction, or (iii) physical and chemical extraction which results in the selective extraction of the pigment relative to the nutritive or aromatic constituent, but does not include (aa) any food (including one in dried or concentrated form), (bb) any flavouring, used, in either case, in the manufacture of any compound food because of its aromatic, sapid or nutritive properties albeit that it may also be used secondarily to add or restore colouring to such compound food, and (cc) any colour when it is used only for colouring any inedible external part of a food.

Processed (in relation to any food):   having undergone any treatment resulting in a substantial change in the original state of the food, but does not include dividing, parting, severing, boning, mincing, skinning, paring, peeling, grinding, cutting, cleaning, trimming, deep-freezing, freezing, chilling, milling, husking, packing or unpacking, and 'unprocessed' shall be construed accordingly.

Permitted Colours

The following are permitted colours (when they satisfy the purity criteria set out in Directive 95/45):

Table 4.2.1

EC No.	Common name
E 100	Curcumin
E 101	(i) Riboflavin
	(ii) Riboflavin-5'-phosphate
E 102	Tartrazine
E 104	Quinoline Yellow
E 110	Sunset Yellow FCF
	Orange Yellow S
E 120	Cochineal, carminic acid, carmines

4.2

(continued)

EC No.	Common name
E 122	Azorubine, carmoisine
E 123	Amaranth
E 124	Ponceau 4R, Cochineal Red A
E 127	Erythrosine
E 128	Red 2G
E 129	Allura Red AC
E 131	Patent Blue V
E 132	Indigotine, Indigo carmine
E 133	Brilliant Blue FCF
E 140	Chlorophylls and chlorophyllins:
	(i) Chlorophylls
	(ii) Chlorophyllins
E 141	Copper complexes of chlorophylls and chlorophyllins
	(i) Copper complexes of chlorophylls
	(ii) Copper complexes of chlorophyllins
E 142	Green S
E 150a	Plain caramel
E 150b	Caustic sulphite caramel
E 150c	Ammonia caramel
E 150d	Sulphite ammonia caramel
E 151	Brilliant Black BN, Black PN
E 153	Vegetable carbon
E 154	Brown FK
E 155	Brown HT
E 160a	Carotenes:
	(i) Mixed carotenes
	(ii) β-Carotene
E 160b	Annatto, bixin, norbixin
E 160c	Paprika extract, capsanthin, capsorubin
E 160d	Lycopene
E 160e	β-apo-β′-Carotenal (C 30)
E 160f	Ethyl ester of β-apo-8′-carotenic acid (C 30)
E 161b	Lutein
E 161g	Canthaxanthin
E 162	Beetroot Red, betanin
E 163	Anthocyanins
E 170	Calcium carbonate
E 171	Titanium dioxide
E 172	Iron oxides and hydroxides
E 173	Aluminium
E 174	Silver
E 175	Gold
E 180	Litholrubine BK

Foods which may not contain added colours

The following foods may not contain added colour except where specifically permitted in the following section or where its presence is as a result of the carryover of colouring legitimately used in ingredients of the foods:

Table 4.2.2

Unprocessed foodstuffs
All bottled or packed waters
Milk, semi-skimmed and skimmed milk, pasteurised or sterilised (including UHT
    sterilisation) (unflavoured)
Chocolate milk
Fermented milk (unflavoured)
Preserved milks as mentioned in Directive 76/118/EEC
Buttermilk (unflavoured)
Cream and cream powder (unflavoured)
Oils and fats of animal or vegetable origin
Eggs and egg products as defined in Article 2(1) of Directive 891437/EEC
Flour and other milled products and starches
Bread and similar products
Pasta and gnocchi
Sugar, including all mono- and disaccharides
Tomato paste and canned and bottled tomatoes
Tomato-based sauces
Fruit juice and fruit nectar as mentioned in Directive 93/77 and vegetable juice
Fruit, vegetables (including potatoes) and mushrooms – canned, bottled or dried;
    processed fruit, vegetables (including potatoes) and mushrooms
Extra jam, extra jelly, and chestnut purée as mentioned in Directive 79/693/EEC;
    crème de pruneaux
Fish, molluscs and crustaceans, meat, poultry and game as well as their
    preparations, but not including prepared meals containing these ingredients
Cocoa products and chocolate components in chocolate products as mentioned in
    Directive 73/241/EEC
Roasted coffee, tea, chicory; tea and chicory extracts; tea, plant, fruit and cereal
    preparations for infusions, as well as mixes and instant mixes of these products
Salt, salt substitutes, spices and mixtures of spices
Wine and other products defined by Regulation (EEC) No. 822/87
Korn, Kornbrand, fruit spirit drinks, fruit spirits, ouzo, grappa, tsikoudia from Crete,
    tsipouro from Macedonia, tsipouro from Thessaly, tsipouro from Tyrnavos, eau de
    vie de marc Marque nationale luxembourgeoise, eau de vie de seigle Marque
    nationale luxembourgeoise, London gin, as defined in Regulation (EEC) No.
    1576/89
Sambuca, maraschino and mistra as defined in Regulation (EEC) No. 1180/91
Sangria, clarea and zurra as mentioned in Regulation (EEC) No. 1601/91
Wine vinegar
Foods for infants and young children as mentioned in Directive 89/398/EEC
    including foods for infants and young children not in good health
Honey
Malt and malt products
Ripened and unripened cheese (unflavoured)
Butter from sheep and goats' milk

4.2

Food to which only certain permitted colours may be added

Table 4.2.3

Food	Permitted colour		Maximum level
Malt bread	E 150a	Plain caramel	*quantum satis*
	E 150b	Caustic sulphite caramel	*quantum satis*
	E 150c	Ammonia caramel	*quantum satis*
	E 150d	Sulphite ammonia caramel	*quantum satis*
Beer	E 150a	Plain caramel	*quantum satis*
Cidre bouché	E 150b	Caustic sulphite caramel	*quantum satis*
	E 150c	Ammonia caramel	*quantum satis*
	E 150d	Sulphite ammonia caramel	*quantum satis*
Butter (including reduced-fat butter and concentrated butter)	E 160a	Carotenes	*quantum satis*
Margarine, minarine, other fat emulsions, and fats essentially free from water	E 160a	Carotenes	*quantum satis*
	E 100	Curcumin	*quantum satis*
	E 160b	Annatto, bixin, norbixin	10 mg/kg
Sage Derby cheese	E 140	Chlorophylls, chlorophyllins	*quantum satis*
	E 141	Copper complexes of chlorophylls and chlorophyllins	*quantum satis*
Ripened Orange, Yellow and broken-white cheese; unflavoured processed cheese	E 160a	Carotenes	*quantum satis*
	E 160c	Paprika extract, capsanthin, capsorubin	*quantum satis*
	E 160b	Annatto, bixin, norbixin	15 mg/kg
Red Leicester cheese	E 160b	Annatto, bixin, norbixin	50 mg/kg
Mimolette cheese	E 160b	Annatto, bixin, norbixin	35 mg/kg
Morbier cheese	E 153	Vegetable carbon	*quantum satis*
Red marbled cheese	E 120	Cochineal, carminic acid, carmines	125 mg/kg
	E 163	Anthocyanins	*quantum satis*
Vinegar	E 150a	Plain caramel	*quantum satis*
	E 150b	Caustic sulphite caramel	*quantum satis*
	E 150c	Ammonia caramel	*quantum satis*
	E 150d	Sulphite ammonia caramel	*quantum satis*
Whisky, whiskey, grain spirit (other than korn or kornbrand or eau de vie de seigle Marque nationale luxembourgeoise), wine spirit, rum, brandy, weinbrand, grape marc, grape marc spirit (other than tsikoudia and tsipouro and eau de vie de marc Marque nationale luxembourgeoise), grappa invecchiata, bagaceira velha as mentioned in Regulation (EEC) No. 1576/89	E 150a	Plain caramel	*quantum satis*
	E 150b	Caustic sulphite caramel	*quantum satis*
	E 150c	Ammonia caramel	*quantum satis*
	E 150d	Sulphite ammonia caramel	*quantum satis*

4.2

(*continued*)

Food	Permitted colour		Maximum level
Aromatized wine-based drinks (except bitter soda) and aromatized wines as mentioned in Regulation (EEC) No. 1601/91	E 150a	Plain caramel	*quantum satis*
	E 150b	Caustic sulphite caramel	*quantum satis*
	E 150c	Ammonia caramel	*quantum satis*
	E 150d	Sulphite ammonia caramel	*quantum satis*
Americano	E 150a	Plain caramel	*quantum satis*
	E 150b	Caustic sulphite caramel	*quantum satis*
	E 150c	Ammonia caramel	*quantum satis*
	E 150d	Sulphite ammonia caramel	*quantum satis*
	E 163	Anthocyanins	*quantum satis*
	E 100	Curcumin	
	E 101	(i) Riboflavin	
		(ii) Riboflavin-5'-phosphate	
	E 102	Tartrazine	100 mg/l
	E 104	Quinoline Yellow	(individually
	E 120	Cochineal, carminic acid, carmines	or in combination)
	E 122	Azorubine, carmoisine	
	E 123	Amaranth	
	E 124	Ponceau 4R	
Bitter soda, bitter vino as mentioned in Regulation (ECC) No. 1601/91	E 150a	Plain caramel	*quantum satis*
	E 150b	Caustic sulphite caramel	*quantum satis*
	E l50c	Ammonia caramel	*quantum satis*
	E 150d	Sulphite ammonia caramel	*quantum satis*
	E 100	Curcumin	
	E 101	(i) Riboflavin	
		(ii) Riboflavin-5'-phosphate	
	E 102	Tartrazine	
	E 104	Quinoline Yellow	
	E 110	Sunset Yellow FCF Orange Yellow S	100 mg/l (individually
	E 120	Cochineal, carminic acid, carmines	or in combination
	E 122	Azorubine, carmoisine	
	E 123	Amaranth	
	E 124	Ponceau 4R, Cochineal Red A	
	E 129	Allura Red AC	
Liqueur wines and quality liqueur wines produced in specified regions	E 150a	Plain caramel	*quantum satis*
	E 150b	Caustic sulphite caramel	*quantum satis*
	E 150c	Ammonia caramel	*quantum satis*
	E 150d	Sulphite ammonia caramel	*quantum satis*
Vegetables in vinegar, brine or oil (excluding olives)	E 101	(i) Riboflavin	*quantum satis*
		(ii) Riboflavin-5'-phosphate	*quantum satis*
	E 140	Chlorophylls, Chlorophyllins	*quantum satis*
	E 150a	Plain caramel	*quantum satis*
	E 150b	Caustic sulphite caramel	*quantum satis*
	E 150c	Ammonia caramel	*quantum satis*
	E 150d	Sulphite ammonia caramel	*quantum satis*

4.2 (*continued*)

Food	Permitted colour		Maximum level
	E 141	Copper complexes of chlorophylls and chlorophyllins	*quantum satis*
	E 160a	Carotenes:	
		(i)i Mixed carotenes	*quantum satis*
		(ii) β-Carotene	*quantum satis*
	E 162	Beetroot Red, betanin	*quantum satis*
	E 163	Anthocyanins	*quantum satis*
Extruded, puffed and/or fruit-flavoured breakfast cereals	E 150c	Ammonia caramel	*quantum satis*
	E 160a	Carotenes	*quantum satis*
	E 160b	Annatto, bixin, norbixin	25 mg/kg
	E 160c	Paprika extract, capsanthin, Capsorubin	*quantum satis*
Fruit-flavoured breakfast cereals	E 120	Cochineal, carminic acid, carmines	200 mg/kg (individually or in combination)
	E 162	Beetroot Red, betanin	
	E 163	Anthocyanins	
Jam, jellies and marmalades as mentioned in Directive 79/693/EEC and other similar fruit preparations including low calorie products	E 100	Curcumin	*quantum satis*
	E 140	Chlorophylls and chlorophyllins	*quantum satis*
	E 141	Copper complexes of chlorophylls and chlorophyllins	*quantum satis*
	E 150a	Plain caramel	*quantum satis*
	E 150b	Caustic sulphite caramel	*quantum satis*
	E 150c	Ammonia caramel	*quantum satis*
	E 150d	Sulphite ammonia caramel	*quantum satis*
	E 160a	Carotenes:	
		(i) Mixed carotenes	*quantum satis*
		(ii) β-Carotene	*quantum satis*
	E 160c	Paprika extract, capsanthin, capsorubin	*quantum satis*
	E 162	Beetroot Red, betanin	*quantum satis*
	E 163	Anthocyanins	*quantum satis*
	E 104	Quinoline Yellow	100 mg/kg (individually or in combination)
	E 110	Sunset Yellow FCF	
	E 120	Cochineal, carminic acid, carmines	
	E 124	Ponceau 4R, Cochineal Red A	
	E 142	Green S	
	E 160d	Lycopene	
	E 161b	Lutein	
Sausages, pâtés and terrines	E 100	Curcumin	20 mg/kg
	E 120	Cochineal, carminic acid, carmines	100 mg/kg
	E 150a	Plain caramel	*quantum satis*
	E 150b	Caustic sulphite caramel	*quantum satis*
	E 150c	Ammonia caramel	*quantum satis*
	E 150d	Sulphite ammonia caramel	*quantum satis*
	E 160a	Carotenes	20 mg/kg

(continued)

4.2

Food	Permitted colour		Maximum level
	E 160c	Paprika extract, capsanthin, capsorubin	10 mg/kg
	E 162	Beetroot Red, betanin	*quantum satis*
Luncheon meat	E 129	Allura Red	25 mg/kg
Breakfast sausages with a minimum cereal content of 6 % burger meat with a minimum vegetable and/ or cereal content of 4%	E 129	Allura Red AC	25 mg/kg
	E 120	Cochineal, carminic acid, carmines	100 mg/kg
	E 128	Red 2G	20 mg/kg
	E 150a	Plain caramel	*quantum satis*
	E 150b	Caustic sulphite caramel	*quantum satis*
	E 150c	Ammonia caramel	*quantum satis*
	E 150d	Sulphite ammonia caramel	*quantum satis*
Chorizo sausage Salchichon	E 120	Cochineal, carminic acid, carmines	200 mg/kg
	E 124	Ponceau 4R, Cochineal Red A	250 mg/kg
Sobrasada	E 110	Sunset Yellow FCF	135 mg/kg
	E 124	Ponceau 4R, Cochineal Red A	200 mg/kg
Pasturmas (edible external coating)	E 100	Curcumin	*quantum satis*
	E 101	(i) Riboflavin	*quantum satis*
		(ii) Riboflavin-5'-phosphate	*quantum satis*
	E 120	Cochineal, carminic acid, carmines	*quantum satis*
Dried potato granules and flakes	E 100	Curcumin	*quantum satis*
Processed mushy and garden peas (canned)	E 102	Tartrazine	100 mg/kg
	E 133	Brilliant Blue FCF	20 mg/kg
	E 142	Green S	10 mg/kg

Colours permitted for use in certain foods only

Table 4.2.4

Permitted colour	Food	Maximum level
E 123 Amaranth	Americano	100 mg/l (but see the entry for Americano in Table 4.2.3)
	Aperitif wines, spirit drinks including products with less than 15% alcohol by volume	30 mg/l
	Bitter soda, bitter vino as mentioned in Regulation (EEC) No. 1601/91	100 mg/l (but see the entry for bitter soda, bitter vino in Table 4.2.3)
	Fish roe	30 mg/kg

4.2

(continued)

Permitted colour		Food	Maximum level
E 127	Erythrosine	Cocktail cherries and candied cherries	200 mg/kg
		Bigarreaux cherries in syrup and in cocktails	150 mg/kg
E 128	Red 2G	Breakfast sausages with a minimum cereal content of 6%	20 mg/kg
		Burger meat with a minimum vegetable and/or cereal content of 4%	
E 154	Brown FK	Kippers	20 mg/kg
E 161g	Canthaxanthin	Saucisses de Strasbourg	15 mg/kg
E 173	Aluminium	External coating of sugar confectionery for the decoration of cakes and pastries	*quantam satis*
E 174	Silver	External coating of confectionery	*quantum satis*
		Decoration of chocolates	*quantum satis*
		Liqueurs	*quantum satis*
E 175	Gold	External coating of confectionery	*quantum satis*
		Decoration of chocolates	*quantum satis*
		Liqueurs	*quantum satis*
E 180	Litholrubine BK	Edible cheese rind	*quantum satis*
E 160b	Annatto, bixin, norbixin	Margarine, minarine, other fat emulsions, and fats essentially free from water	10 mg/kg
		Decorations and coatings	20 mg/kg
		Fine bakery wares	10 mg/kg
		Edible ices	20 mg/kg
		Liqueurs, including fortified beverages with less than 15% alcohol by volume	10 mg/l
		Flavoured processed cheese, Ripened Orange, Yellow and broken-white cheese, unflavoured processed cheese	15 mg/kg
		Desserts	10 mg/kg
		'Snacks': dry, savoury potato, cereal or starch-based snack products:	
		– Extruded or expanded savoury snack products	20 mg/kg
		– Other savoury snack products and savoury coated nuts	10 mg/kg

4.2

Permitted colour	Food	Maximum level
	Smoked fish	10 mg/kg
	Edible cheese rind and edible casings	20 mg/kg
	Red Leicester cheese	50 mg/kg
	Mimolette cheese	35 mg/kg
	Extruded, puffed and/or fruit-flavoured breakfast cereals	25 mg/kg

Further permitted colours in certain foods

1) The following colours may be used, in each case at *quantum satis*, in foods listed in Table 4.2.7 below and in any food other than one listed in Tables 4.2.2 and 4.2.3.

Table 4.2.5

EC No. and name
E 101   (i) Riboflavin
(ii) Riboflavin-5′-phosphate
E 140   Chlorophylls and chlorophyllins
E 141   Copper complexes of chlorophylls and chlorophyllins
E 150a   Plain caramel
E 150b   Caustic sulphite caramel
E 150c   Ammonia caramel
E 150d   Sulphite ammonia caramel
E 153   Vegetable carbon
E 160a   Carotenes
E 160c   Paprika extract, capsanthin, capsorubin
E 162   Beetroot Red, betanin
E 163   Anthocyanins
E 170   Calcium carbonate
E 171   Titanium dioxide
E 172   Iron oxides and hydroxides

2) The following colours may be used singly or in combination in the foods listed in Table 4.2.7, in each case up to the maximum level specified for such food. However, for non-alcoholic flavoured drinks, edible ices, desserts, fine bakery wares and confectionery, the colours may be used up to the limit indicated in Table 4.2.7 but the quantities of each of the colours E 110, E 122, E 124 and E 155 may not exceed 50 mg/kg or mg/l.

4.2

Table 4.2.6

EC No. and name	
E 100	Curcumin
E 102	Tartrazine
E 104	Quinoline Yellow
E 110	Sunset Yellow FCF
	Orange Yellow S
E 120	Cochineal, carminic acid, carmines
E 122	Azorubine, carmoisine
E 124	Ponceau 4R, Cochineal Red A
E 129	Allura Red AC
E 131	Patent Blue V
E 132	Indigotine, Indigo carmine
E 133	Brilliant Blue FCF
E 142	Green S
E 151	Brilliant Black BN, Black PN
E 155	Brown HT
E 160d	Lycopene
E 160e	β-apo-8′-Carotenal (C 30)
E 160f	Ethyl ester of β-apo-8′-carotenic acid (C 30)
E 161b	Lutein

Table 4.2.7

Food	Maximum level
Non-alcoholic flavoured drinks	100 mg/l
Candied fruits and vegetables, mostarda di frutta	200 mg/kg
Preserves of red fruits	200 mg/kg
Confectionery	300 mg/kg
Decorations and coatings	500 mg/kg
Fine bakery wares (e.g. viennoiserie, biscuits, cakes and wafers)	200 mg/kg
Edible ices	150 mg/kg
Flavoured processed cheese	100 mg/kg
Desserts including flavoured milk products	150 mg/kg
Sauces, seasonings (for example, curry powder, tandoori), pickles, relishes, chutney and piccalilli	500 mg/kg
Mustard	300 mg/kg
Fish paste and crustacean paste	100 mg/kg
Precooked crustaceans	250 mg/kg
Salmon substitutes	500 mg/kg
Surimi	500 mg/kg
Fish roe	300 mg/kg
Smoked fish	100 mg/kg
'Snacks': dry, savoury potato, cereal or starch-based snack products	
– Extruded or expanded savoury snack products	200 mg/kg
– Other savoury snack products and savoury coated nuts	100 mg/kg
Edible cheese rind and edible casings	quantum satis
Complete formulae for weight control intended to replace total daily food intake or an individual meal	50 mg/kg
Complete formulae and nutritional supplements for use under medical supervision	50 mg/kg

4.2

(continued)

Food	Maximum level
Liquid food supplements/dietary integrators	100 mg/l
Solid food supplements/dietary integrators	300 mg/kg
Soups	50 mg/kg
Meat and fish analogues based on vegetable proteins	100 mg/kg
Spiritous beverages (including products less than 15% alcohol by volume), except any mentioned in Tables 4.2.2 or 4.2.3	200 mg/l
Aromatized wines, aromatized wine-based drinks and aromatized wine-product cocktails as mentioned in Regulation (EEC) No. 1601/91, except any mentioned in Tables 4.2.2 or 4.2.3	200 mg/l
Fruit wines (still or sparkling); cider (except cidre bouché) and perry; aromatized fruit wines, cider and perry	200 mg/l

Flavourings

Regulation: Flavourings in Food Regulations 1992 (1992/1971)
Amendment: Flavourings in Food (Amendment) Regulations 1994 (1994/1486)
Food Labelling Regulations 1996 (1996/1499)

Definitions

Flavouring: (as a noun) material used or intended for use in or on food to impart odour, taste or both (as an adjective); flavouring shall be construed accordingly.

Relevant flavouring: flavouring which does not consist entirely of excepted material and the components of which include at least one of the following: (i) a flavouring substance, (ii) a flavouring preparation, (iii) a process flavouring, (iv) a smoke flavouring.

Flavouring substance: a chemical substance with flavouring properties the chemical structure of which has been established by methods normally used among scientists and which is:

(i) obtained by physical (including distillation and solvent extraction), enzymatic or microbiological processes from appropriate material of vegetable or animal origin (termed 'natural flavouring substances' for labelling for business sales);

(ii) either obtained by chemical synthesis or isolated by chemical processes and which is chemically identical to a substance naturally present in appropriate material of vegetable or animal origin (termed 'flavouring substances identical to natural substances' for labelling for business sales); or

(iii) obtained by chemical synthesis but not included under (ii) (termed 'artificial flavouring substances' for labelling for business sales).

Flavouring preparation: a product (other than a flavouring substance), whether concentrated or not, with flavouring properties which is obtained by physical processes (including distillation and solvent extraction) or by enzymatic or microbiological processes from appropriate material of vegetable or animal origin.

4.2

Process flavouring: a product which is obtained according to good manufacturing practices by heating to a temperature not exceeding 180°C for a continuous period not exceeding 15 minutes a mixture of ingredients (whether or not with flavouring properties) of which at least one contains nitrogen (amino) and another is a reducing sugar.

Smoke flavouring: an extract from smoke of a type normally used in food smoking processes.

Permitted flavouring: relevant flavouring which:
- (i) has in it or on it no specified substance which has been added as such;
- (ii) contains no element or substance in a toxicologically dangerous quantity, and
- (iii) contains less than 3 mg/kg arsenic, 10 mg/kg lead, 1 mg/kg cadmium and 1 mg/kg mercury.

Excepted material: any edible substance (including herbs and spices) or product, intended for human consumption as such, with or without reconstitution, and any substance which has exclusively a sweet, sour or salt taste.

Specified substance: those substances given in the table below.

Appropriate material of vegetable or animal origin: material of vegetable or animal origin which is either raw or has been subjected to a process (including drying, torrefaction and fermentation) normally used in preparing food for human consumption and to no process other than one normally so used.

Specified substances

The following substances are specified and
- a) must not be added to permitted flavourings (see definition above);
- b) no food sold which has in it or on it any relevant flavouring shall contain any specified substance other than a specified substance which is present in the food naturally or as a result of the inclusion of the relevant flavouring where that relevant flavouring is prepared from natural raw materials;
- c) no food sold which has in it or on it any relevant flavouring shall contain a proportion greater than that specified in the table except where the proportion of the specified substance in the food is not greater than it would have been had the food not had the relevant flavouring in it or on it.

Specified substance	Standard permitted proportion (mg/kg)	Permitted proportion in the case of particular descriptions of food
Agaric acid	20	(a) Alcoholic drinks: 100 mg/kg (b) Food (other than drinks) containing mushrooms: 100 mg/kg
Aloin	0.1	Alcoholic drinks: 50 mg/kg

(continued)

Specified substance	Standard permitted proportion (mg/kg)	Permitted proportion in the case of particular descriptions of food
β Asarone	0.1	(a) Alcoholic drinks: 1 mg/kg (b) Seasonings used in snack foods: 1 mg/kg
Berberine	0.1	Alcoholic drinks: 10 mg/kg
Coumarin	2	(a) Chewing gum: 50 mg/kg (b) Alcoholic drinks: 10 mg/kg (c) Caramel confectionery: 10 mg/kg
Hydrocyanic acid	1	(a) Nougat, marzipan, a nougat or marzipan substitute or similar product: 50 mg/kg (b) Tinned stone fruit: 5 mg/kg (c) Alcoholic drinks: 1 mg/kg for each percentage of alcohol by volume therein
Hypericine	0.1	(a) Alcoholic drinks: 10 mg/kg (b) Confectionery: 1 mg/kg
Pulegone	25	(a) Mint confectionery: 350 mg/kg (b) Mint or peppermint flavoured drinks: 250 mg/kg (c) Other drinks: 100 mg/kg
Quassine	5	(a) Alcoholic drinks: 50 mg/kg (b) Confectionery in pastille form: 10 mg/kg
Safrole, isosafrole any combination of these	1	(a) Food containing mace, nutmeg or both: 15 mg/kg (b) Alcoholic drinks containing more than 25% alcohol by volume: 5 mg/kg (c) Other alcoholic drinks: 2 mg/kg
Santonin	0.1	Alcoholic drinks containing more than 25% alcohol by volume: 1 mg/kg
Thuyone α–, thuyone β– or any combination of these	0.5	(a) Bitters: 35 mg/kg (b) Food (other than drinks) containing preparations based on sage: 25 mg/kg (c) Alcoholic drinks containing more than 25% alcohol by volume: 10 mg/kg (d) Other alcoholic drinks: 5 mg/kg

General points

1) Any relevant flavourings added to food must be permitted flavourings.
2) Any relevant flavouring added to food must not increase the proportion of benzo[3,4]pyrene by more than 0.03 g per kilogram of food above that which would otherwise occur in the food.
3) The Regulations contain detailed labelling provision (see Regulations). These are more demanding for business sales which include a requirement to specify, in descending weight order, the

4.2

components of relevant flavourings using the following classification (see definitions above): 'natural flavouring substances', 'flavouring substances identical to natural substances', 'artificial flavouring substances', 'flavouring preparations', 'process flavourings' and 'smoke flavourings' and for each other substance, its name or 'E' number.

4) Relevant flavouring may not be described as 'natural' in any business or consumer sale unless it is either in compliance with (3), or that relevant flavouring is a permitted flavouring the flavouring components of which are exclusively composed of
   a) flavouring substances which come within the category of 'natural flavouring substances',
   b) flavouring preparations, or
   c) both.

5) In a business or consumer sale of relevant flavouring, the word 'natural' may not be used to qualify any substance used in its preparation unless that relevant flavouring is a permitted flavouring the flavouring component of which has been isolated solely, or almost solely, from that substance by physical processes, enzymatic or microbiological processes or processes normally used in preparing food for human consumption.

Mineral hydrocarbons

Regulation: Mineral Hydrocarbons in Food Regulations 1966 (1966/1073)
Amendment: Miscellaneous Food Additives Regulations 1995 (1995/3187)

Definition
Mineral hydrocarbon: any hydrocarbon product, whether liquid, semi-liquid or solid, derived from any substance of mineral origin and includes liquid paraffin, white oil, petroleum jelly, hard paraffin and microcrystalline wax.

Requirements
1) Except for those cases listed in (2), no person shall
   a) use or permit to be used any mineral hydrocarbon in the composition or preparation of any food; or
   b) sell, consign or deliver, or import any food containing any mineral hydrocarbon.

2) The following exemptions are provided from the general prohibition established by (1) provided that the mineral hydrocarbon meets the required specification (see Regulations):
   a) foods containing mineral hydrocarbon because of the use of mineral hydrocarbon as a lubricant or greasing agent on some surface with which it has necessarily come into contact, maximum 0.2%;

b) chewing compound, maximum 60% solid mineral hydrocarbon
   (with a separate special specification);
c) rind of whole pressed cheese (no limit specified);
d) any food containing mineral hydrocarbon as permitted by the
   Miscellaneous Food Additives Regulations and subject to the
   requirements of those Regulations.

4.2

Miscellaneous additives

Regulation: Miscellaneous Food Additives Regulations 1995 (1995/3187)

Definitions (general)

Miscellaneous additive: any food additive which is used or intended to
be used primarily as an acid, acidity regulator, anti-caking agent, anti-
foaming agent, antioxidant, bulking agent, carrier, carrier solvent,
emulsifier, emulsifying salt, firming agent, flavour enhancer, foaming
agent, gelling agent, glazing agent, humectant, modified starch,
packaging gas, preservative, propellant, raising agent, sequestrant,
stabiliser or thickener, but does not include any processing aid.

Food additive:

a) any substance not normally consumed as a food in itself and not
   normally used as a characteristics ingredient of food, whether or not
   it has nutritive value, the intentional addition of which to food for a
   technological purpose in the manufacture, processing, preparation,
   treatment, packaging, transport or storage of such food results, or
   may reasonably be expected to result, in it or its byproducts
   becoming directly or indirectly a component of such foods; or

b) a carrier or carrier solvent: but does not include:

   (i)   any substance used for the treatment of drinking water (as
           permitted by EC Directives);
   (ii)   any product containing pectin and derived from dried apple
           pomace or peel of citrus fruit, or from a mixture of both, by the
           action of dilute acid followed by partial neutralisation with
           sodium or potassium salts (liquid pectin);
   (iii)   chewing-gum bases;
   (iv)   white or yellow dextrin, roasted or dextrinated starch, starch
           modified by acid or alkali treatment, bleached starch,
           physically modified starch and starch treated with amylolitic
           enzymes;
   (v)   ammonium chloride;
   (vi)   blood plasma, edible gelatin, protein hydrolysates and their
           salts, milk protein and gluten;
   (vii)   amino acids and their salts (other than glumatic acid, glycine,
           cysteine and cystine and their salts) having no additive
           function;

4.2

    (viii)   casein and caseinates;
    (ix)   inulin.

Processed: in relation to any food, means having undergone any treatment resulting in a substantial change in the original state of the food, but does not include dividing, parting, severing, boning, mincing, skinning, paring, peeling, grinding, cutting, cleaning, trimming, deep-freezing, freezing, chilling, milling, husking, packing, or unpacking; and 'unprocessed' shall be construed accordingly.

Processing aid: any substance not consumed as a food by itself, intentionally used in the processing of raw materials, foods or their ingredients to fulfil certain technological purposes during treatment or processing, and which may result in the unintentional but technically unavoidable presence of residues of the substance or its derivatives in the final product, provided that these residues do not present any health risk and do not have any technological effect on the finished product.

Quantum satis: no maximum level of permitted miscellaneous additive is specified but it may be used in accordance with good manufacturing practice at a level not higher than is necessary to achieve the intended purpose and provided that such use does not mislead the consumer.

Definitions (additive categories)

Acid: any substance which increases the acidity of a food or imparts a sour taste to it, or both.

Acidity regulator: any substance which alters or controls the acidity or alkalinity of a food.

Anti-caking agent: any substance which reduces the tendency of individual particles of a food to adhere to one another.

Anti-foaming agent: any substance which prevents or reduces foaming.

Antioxidant: any substance which prolongs the shelf-life of a food by protecting it against deterioration caused by oxidation, including fat rancidity and colour changes.

Bulking agent: any substance which contributes to the volume of a food without contributing significantly to its available energy value.

Carrier/carrier solvent: any substance, other than a substance generally considered as food, used to dissolve, dilute, disperse or otherwise physically modify a miscellaneous additive, colour or sweetener, or an enzyme which is acting as a processing aid, without altering its technological function (and without exerting any technological effect itself) in order to facilitate its handling, application or use.

Emulsifier: any substance which makes it possible to form or maintain a homogeneous mixture of two or more immiscible phases, such as oil and water, in a food.

Emulsifying salt: any substance which converts proteins contained in cheese into a dispersed form, thereby bringing about homogeneous distribution of fat and other components.

4.2

Firming agent: any substance which makes or keeps tissues of fruit or vegetables firm or crisp or which interacts with a gelling agent to produce or strengthen a gel.

Flavour enhancer: any substance which enhances the existing taste or odour, or both, of a food.

Foaming agent: any substance which makes it possible to form a homogeneous dispersion of a gaseous phase in a liquid or solid food.

Gelling agent: any substance which gives a food texture through the formation of a gel.

Glazing agent: any substance which, when applied to the external surface of a food, imparts a shiny appearance or provides a protective coating, and includes lubricants.

Humectant: any substance which prevents a food from drying out by counteracting the effect of an atmosphere having a low degree of humidity, or which promotes the dissolution of a powder in an aqueous medium.

Modified starch: any substance obtained by one or more chemical treatments of edible starch, which may have undergone a physical or enzymatic treatment, and may be acid or alkali thinned or bleached.

Packaging gas: any gas, other than air, which is introduced into a container before, during or after the placing of a food in that container.

Preservative: any substance which prolongs the shelf-life of a food by protecting it against deterioration caused by micro-organisms.

Propellant: any gas, other than air, which expels a food from a container.

Raising agent: any substance or a combination of substances which liberates gas and thereby increases the volume of a dough or a batter.

Sequestrant: any substance which forms a chemical complex with metallic ions.

Stabiliser: any substance which makes it possible to maintain the physico-chemical state of a food, including any substance which enables a homogeneous dispersion of two or more immiscible substances in a food to be maintained, and any substance which stabilises, retains or intensifies an existing colour of a food.

Thickener: any substance which increases the viscosity of a food.

Miscellaneous additives generally permitted for use in food
The following additives may be used generally in foods unless the foods are subject to the restrictions listed (see Tables 4.2.18 to 4.2.22).

4.2

Table 4.2.8

EC No.	and name	EC No.	and name
E 170	Calcium carbonates: (i) Calcium carbonate (ii) Calcium hydrogen carbonate	E 352	Calcium malates: (i) Calcium malate (ii) Calcium hydrogen malate
E 260	Acetic acid	E 354	Calcium tartrate
E 261	Potassium acetate	E 380	Triammonium citrate
E 262	Sodium acetates: (i) Sodium acetate (ii) Sodium hydrogen acetate (sodium diacetate)	E 400	Alginic acid
		E 401	Sodium alginate
		E 402	Potassium alginate
		E 403	Ammonium alginate
E 263	Calcium acetate	E 404	Calcium alginate
E 270	Lactic acid	E 406	Agar
E 290	Carbon dioxide (2)	E 407	Carrageenan (1)
E 296	Malic acid	E 410	Locust bean gum (3)
E 300	Ascorbic acid	E 412	Guar gum (3)
E 301	Sodium ascorbate	E 413	Tragacanth
E 302	Calcium ascorbate	E 414	Acacia gum (gum arabic)
E 304	Fatty acid esters of ascorbic acid: (i) Ascorbyl palmitate (ii) Ascorbyl stearate	E 415	Xanthan gum (3)
		E 417	Tara gum (3)
		E 418	Gellan gum
E 306	Tocopherol-rich extract	E 422	Glycerol
E307	α-Tocopherol	E 440	Pectins (1): (i) Pectin (ii) Amidated pectin
E 308	γ-Tocopherol		
E 309	δ-Tocopherol		
E 322	Lecithins	E 460	Cellulose: (i) Microcrystalline cellulose (ii) Powdered cellulose
E 325	Sodium lactate		
E 326	Potassium lactate		
E 327	Calcium lactate	E 461	Methyl cellulose
E 330	Citric acid	E 463	Hydroxypropyl cellulose
E 331	Sodium citrates: (i) Monosodium citrate (ii) Disodium citrate (iii) Trisodium citrate	E 464	Hydroxypropyl methyl cellulose
		E 465	Ethyl methyl cellulose
		E 466	Carboxy methyl cellulose Sodium carboxy methyl cellulose
E 332	Potassium citrates: (i) Monopotassium citrate (ii) Tripotassium citrate	E 470a	Sodium, potassium and calcium salts of fatty acids
E 333	Calcium citrates: (i) Monocalcium citrate (ii) Dicalcium citrate (iii) Tricalcium citrate	E 470b	Magnesium salts of fatty acids
		E 471	Mono- and diglycerides of fatty acids
		E 472a	Acetic acid esters of mono- and diglycerides of fatty acids
E 334	Tartaric acid (L(+)-)	E 472b	Lactic acid esters of mono- and diglycerides of fatty acids
E 335	Sodium tartrates: (i) Monosodium tartrate (ii) Disodium tartrate	E 472c	Citric acid esters of mono- and diglycerides of fatty acids
E 336	Potassium tartrates: (i) Monopotassium tartrate (ii) Dipotassium tartrate	E 472d	Tartaric acid esters of mono- and diglycerides of fatty acids
E 337	Sodium potassium tartrate	E 472e	Mono- and diacetyl tartaric acid esters of mono- and diglycerides of fatty acids
E 350	Sodium malates: (i) Sodium malate (ii) Sodium hydrogen malate		
E 351	Potassium malate		

(continued)

4.2

EC No.	and name	EC No.	and name
E 472f	Mixed acetic and tartaric acid esters of mono- and diglycerides of fatty acids	E 516	Calcium sulphate
		E 524	Sodium hydroxide
		E 525	Potassium hydroxide
E 500	Sodium carbonates:	E 526	Calcium hydroxide
	(i) Sodium carbonate	E 527	Ammonium hydroxide
	(ii) Sodium hydrogen carbonate	E 528	Magnesium hydroxide
	(iii) Sodium sesquicarbonate	E 529	Calcium oxide
E 501	Potassium carbonates:	E 530	Magnesium oxide
	(i) Potassium carbonate	E 570	Fatty acids
	(ii) Potassium hydrogen carbonate	E 574	Gluconic acid
		E 575	Glucono-δ-lactone
E 503	Ammonium carbonates:	E 576	Sodium gluconate
	(i) Ammonium carbonate	E 577	Potassium gluconate
	(ii) Ammonium hydrogen carbonate	E 578	Calcium gluconate
		E 640	Glycine and its sodium salt
E 504	Magnesium carbonates:	E 938	Argon (2)
	(i) Magnesium carbonate	E 939	Helium (2)
	(ii) Magnesium hydroxide carbonate (syn.: magnesium hydrogen carbonate)	E 941	Nitrogen (2)
		E 942	Nitrous oxide (2)
		E 948	Oxygen (2)
E 507	Hydrochloric acid	E 1200	Polydextrose
E 508	Potassium chloride	E 1404	Oxidised starch
E 509	Calcium chloride	E 1410	Monostarch phosphate
E 511	Magnesium chloride	E l 412	Distarch phosphate
E 513	Sulphuric acid	E 1413	Phosphated distarch phosphate
E 514	Sodium sulphates:	E 1414	Acetylated distarch phosphate
	(i) Sodium sulphate	E 1420	Acetylated starch
	(ii) Sodium hydrogen sulphate	E 1422	Acetylated distarch adipate
E 515	Potassium sulphates:	E 1440	Hydroxy propyl starch
	(i) Potassium sulphate	E 1442	Hydroxy propyl distarch phosphate
	(ii) Potassium hydrogen sulphate	E 1450	Starch sodium octenyl succinate

Notes:

  (1) The substances listed under numbers E 407 and E 440 may be standardised with sugars, on condition that this is stated in addition to the number and designation

  (2) The substances E 290, E 938, E 939, E 941, E 942 and E 948 may also be used at *quantum satis* in the foods referred to in Tables 4.2.18 to 4.2.22

  (3) The substances E 410, E 412, E 415 and E 417 may not be used to produce dehydrated foods intended to rehydrate on ingestion

Conditionally permitted preservatives and antioxidants

  1) The sorbates, benzoates and *p*-hydroxybenzoates listed in Table 4.2.9 may be used in the foods listed in Table 4.2.10 up to the levels specified (foods ready for consumption prepared following manufacturer's instructions).

4.2

Table 4.2.9

EC No. and name		Abbreviation used in Table 4.2.10
E 200	Sorbic acid	Sa
E 202	Potassium sorbate	
E 203	Calcium sorbate	
E 210	Benzoic acid (1)	Ba
E 211	Sodium benzoate	
E 212	Potassium benzoate	
E 213	Calcium benzoate	
E 214	Ethyl-*p*-hydroxybenzoate	PHB
E 215	Sodium ethyl-*p*-hydroxybenzoate	
E 216	Propyl-*p*-hydroxybenzoate	
E 217	Sodium propyl-*p*-hydroxybenzoate	
E 218	Methyl-*p*-hydroxybenzoate	
E 219	Sodium methyl-*p*-hydroxybenzoate	

Note: (1) Benzoic acid may be present in certain fermented products resulting from the fermentation process following good manufacturing practice

Table 4.2.10

Food	Maximum level (mg/kg or mg/l as appropriate)					
	Sa	Ba	PHB	Sa + Ba	Sa + PHB	Sa + Ba + PHB
Wine-based flavoured drinks including products covered by Regulation (EEC) No. 1601/91	200					
Non-alcoholic flavoured drinks (excluding dairy-based drinks)	300	150		250 Sa+ 150 Ba		
Liquid tea concentrates and liquid fruit and herbal infusion concentrates				600		
Grape juice, unfermented for sacramental use				2000		
Wines as referred to in Regulation (EEC) No. 822/87; alcohol-free wine; fruit wine (including alcohol-free); *made wine*; cider and perry (including alcohol-free)	200					
Søl ... Saft or *Sødet ... Saft*	500	200				
Alcohol-free beer in keg		200				
Mead	200					

(continued)

4.2

Food	Maximum level (mg/kg or mg/l as appropriate)					
	Sa	Ba	PHB	Sa + Ba	Sa + PHB	Sa + Ba + PHB
Spirits with less than 15% alcohol by volume	200	200		400		
Fillings of ravioli and similar products	1000					
Low-sugar jams, jellies, marmalades and similar low calorie or sugar-free products and other fruit based spreads; mermeladas		500		1000		
Candied, crystallised and glacé fruit and vegetables				1000		
Dried fruit	1000					
Frugtgrød and Rote Grütze	1000	500				
Fruit and vegetable preparations including fruit-based sauces, excluding purée, mousse, compote, salads and similar products, canned or bottled	1000					
Vegetables in vinegar, brine or oil (excluding olives)				2000		
Potato dough and pre-fried potato slices	2000					
Gnocchi	1000					
Polenta	200					
Olives and olive-based preparations	1000					
Jelly coatings of meat products (cooked, cured or dried); pâté					1000	
Surface treatment of dried meat products						quantum satis
Semi-preserved fish products including fish roe products				2000		
Salted, dried fish				200		
Shrimps, cooked				2000		

4.2

(continued)

Food	Maximum level (mg/kg or mg/l as appropriate)					
	Sa	Ba	PHB	Sa + Ba	Sa + PHB	Sa + Ba + PHB
Crangon crangon and *Crangon vulgaris,* cooked				6000		
Cheese, pre-packed, sliced	1000					
Unripened cheese	1000					
Processed cheese	2000					
Layered cheese and cheese with added foods	1000					
Non-heat-treated dairy-based desserts				300		
Curdled milk	1000					
Liquid egg (white, yolk or whole egg)				5000		
Dehydrated, concentrated, frozen and deep-frozen egg products	1000					
Prepacked sliced bread and rye-bread	2000					
Partially baked, prepacked bakery wares intended for retail sale	2000					
Fine bakery wares with a water activity of more than 0.65	2000					
Cereal- or potato-based snacks and coated nuts					1000 (max. 300 PHB)	
Batters	2000					
Confectionery (excluding chocolate)						1500 (max. 300 PHB)
Chewing gum				1500		
Toppings (syrups for pancakes, flavoured syrup for milkshakes and ice-cream; similar products)	1000					
Fat emulsions (excluding butter) with a fat content of 60% or more	1000					
Fat emulsions with a fat content less than 60%	2000					
Emulsified sauces with a fat content of 60% or more	1000					

(*continued*)

4.2

Food	*Maximum level* (mg/kg or mg/l as appropriate)					
	Sa	*Ba*	*PHB*	*Sa + Ba*	*Sa + PHB*	*Sa + Ba + PHB*
Emulsified sauces with a fat content less than 60%	2000					
Non-emulsified sauces				1000		
Prepared salads				1500		
Mustard				1000		
Seasonings and condiments				1000		
Liquid soups and broths (excluding canned)				500		
Aspic	1000	500				
Liquid dietary food supplements						2000
Dietetic foods intended for special medical purposes excluding foods for infants and young children as referred to in Directive 89/398/EEC – dietetic formulae for weight control intended to replace total daily food intake or an individual meal				1500		

2) Suphur dioxide and the sulphites listed in Table 4.2.11 may be used in the foods listed in Table 4.2.12 up to the levels specified (foods ready for consumption prepared following manufacturer's instructions).

Table 4.2.11

E 220	Sulphur dioxide
E 221	Sodium sulphite
E 222	Sodium hydrogen sulphite
E 223	Sodium metabisulphite
E 224	Potassium metabisulphite
E 226	Calcium sulphite
E 227	Calcium hydrogen sulphite
E 228	Potassium hydrogen sulphite

4.2

Table 4.2.12

Food	Maximum level (mg/kg or mg/l as appropriate), expressed as SO_2 (1, 2)
Burger meat with a minimum vegetable and/or cereal content of 4%	450
Breakfast sausages	450
Longaniza fresca and *Butifarra fresca*	450
Dried salted fish of the 'Gadidae' species	200
Crustaceans and cephalopods	
– fresh, frozen and deep-frozen crustaceans, *penaeidae solenceridae,. aristeidae* family:	150 (3)
– up to 80 units	150 (3)
– between 80 and 120 units	200 (3)
– over 120 units	300 (3)
– cooked	50 (3)
Dry biscuit	50
Starches (excluding starches for weaning foods, follow-on formulae and infant formulae)	50
Sago	30
Pearl barley	30
Dehydrated granulated potatoes	400
Cereal- and potato-based snacks	50
Peeled potatoes	50
Processed potatoes (including frozen and deep-frozen potatoes)	100
Potato dough	100
White vegetables, dried	400
White vegetables, processed (including frozen and deep-frozen white vegetables)	50
Dried ginger	150
Dried tomatoes	200
Horseradish pulp	800
Onion, garlic and shallot pulp	300
Vegetables and fruits in vinegar, oil or brine (except olives and golden peppers in brine)	100
Golden peppers in brine	500
Processed mushrooms (including frozen mushrooms)	50
Dried mushrooms	100
Dried fruits	
– apricots, peaches, grapes, prunes and figs	2000
– bananas	1000
– apples and pears	600
– other (including nuts in shell)	500

(*continued*)

Food	Maximum level (mg/kg or mg/l as appropriate), *expressed as* SO_2 (1, 2)
Dried coconut	50
Candied, crystallised or glacé fruit, vegetables, angelica and citrus peel	100
Jam, jelly and marmalade as defined in Directive 79/693 (except extra jam and extra jelly) and other similar fruit spreads including low-calorie products	50
Jams, jellies and *marmalades* made with sulphited fruit	100
Fruit-based pie fillings	100
Citrus-juice-based seasonings	200
Concentrated grape juice for home wine-making	2000
Mostarda di frutta	100
Jellying fruit extract, liquid pectin for sale to the final consumer	800
Bottled whiteheart cherries, rehydrated dried fruit and lychees	100
Bottled, sliced lemon	250
Sugars as defined in Directive 73/437 except glucose syrup, whether or not dehydrated	15
Glucose syrup, whether or not dehydrated	20
Treacle and molasses	70
Other sugars	40
Toppings (syrups for pancakes, flavoured syrups for milkshakes and ice cream; similar products)	40
Orange, grapefruit, apple and pineapple juice for bulk dispensing in catering establishments	50
Lime and lemon juice	350
Concentrates based on fruit juice and containing not less than 2.5% barley (*barley water*)	350
Other concentrates based on fruit juice or comminuted fruit; *capilé groselha*	250
Non-alcoholic flavoured drinks containing fruit juice	20 (carryover from concentrates only)
Non-alcoholic flavoured drinks containing at least 235 g/l glucose syrup	50
Grape juice, unfermented, for sacramental use	70
Glucose-syrup-based confectionery	50 (carryover from the glucose syrup only)
Beer including low-alcohol and alcohol-free beer	20

4.2

(continued)

Food	Maximum level (mg/kg or mg/l as appropriate), *expressed as SO$_2$ (1, 2)*
Beer with a second fermentation in the cask	50
Wines	in accordance with Regulations (EEC) No. 822/87. (EEC) No. 4252/88 (EEC) No. 2332192 and (EEC) No. 1873/84 and their implementing regulations; *(pro memoria)* in accordance with Regulation (EEC) No. 1873/84 authorising the offer or disposal for direct human consumption of certain imported wines which may have undergone oenological processes not provided for in Regulation (EEC) No. 337/79
Alcohol-free wine	200
Made wine	260
Cider, perry, fruit wine, sparkling fruit wine (including alcohol-free products)	200
Mead	200
Fermentation vinegar	170
Mustard, excluding Dijon mustard	250
Dijon mustard	500
Gelatin	50
Vegetable- and cereal-protein-based meat, fish and crustacean analogues	200

Notes:
  (1) Maximum levels are expressed as SO$_2$ in mg/kg or mg/l as appropriate and relate to the total quantity, available from all sources
  (2) An SO$_2$ content of not more than 10 mg/kg or 10 mg/l is not considered to be present.
  (3) In edible parts

  3) The preservatives listed in Tables 4.2.13 and 4.2.14 may only be used as specified in the tables and only up to the maximum level indicated.

Table 4.2.13

EC No. and name	Food	Maximum level
E 230 Biphenyl, diphenyl	Surface treatment of citrus fruits	70 mg/kg
E 231 Orthophenyl phenol E 232 Sodium orthophenyl phenol	Surface treatment of citrus fruits	12 mg/kg individually or in combination, expressed as *o*-phenyl phenol
E 233 Thiabendazole	Surface treatment of: – citrus fruit – bananas	 6 mg/kg 3 mg/kg
E 234 Nisin (1)	Semolina and tapioca puddings and similar products	3 mg/kg
	Ripened cheese and processed cheese	12.5 mg/kg
	Clotted cream	10 mg/kg
E 235 Natamycin	Surface treatment of: – hard, semi-hard and semi-soft cheese – dried, cured sausages	} 1 mg/dm² surface (not present at a depth of 5 mm)
E 239 Hexamethylene tetramine	*Provolone* cheese	25 mg/kg residual amount, expressed as formaldehyde
E 242 Dimethyl dicarbonate	Non-alcoholic flavoured drinks Alcohol-free wine Liquid-tea concentrate	} 250 mg/l ingoing amount, residues not detectable
E 284 Boric acid E 285 Sodium tetraborate (borax)	} Sturgeons' eggs (caviar)	4 g/kg expressed as boric acid

Note: (1) This substance may be present naturally in certain cheeses as a result of fermentation processes

Table 4.2.14

EC No. and name	Food	Indicative ingoing amount (mg/kg)	Residual amount (mg/kg)
E 249 Potassium nitrite (1) E 250 Sodium nitrite (1)	Non-heat-treated, cured, dried meat products	150 (2)	50 (3)
	Other cured meat products Canned meat products *Foie gras, foie gras entier, blocs de foie gras*	150 (2)	100 (3)
	Cured bacon		175 (3)

101

4.2 (*continued*)

EC No. and name		Food	Indicative ingoing amount (mg/kg)	Residual amount (mg/kg)
E 251	Sodium nitrate	Cured meat products	300	250 (4)
E 252	Potassium nitrate	Canned meat products		
		Hard, semi-hard and semi-soft cheese Dairy-based cheese analogue		50 (4)
		Pickled herring and sprat		200 (5)

EC No. and name		Food	Maximum level (mg/kg)
E 280	Propionic acid (6)	Prepacked sliced bread and rye bread	3000 (expressed as propionic acid)
E 281	Sodium proprionate (6)		
E 282	Calcium proprionate (6)	Energy reduced bread	
E 283	Potassium propionate (6)	Partially baked, pre-packed bread Prepacked fine bakery wares (including flour confectionery) with a water activity of more than 0.65 Prepacked rolls, buns and *pittas*	2000 (as expressed propionic acid)
		Christmas pudding Prepacked bread	1000 (expressed as propionic acid)
E 1105	Lysozyme	Ripened cheese	*quantum satis*

Notes:
  (1) When labelled 'for food use', nitrite may only be sold in a mixture with salt or a salt substitute
  (2) Expressed as $NaNO_2$
  (3) Residual amount at point of sale to the final consumer, expressed as $NaNO_2$
  (4) Expressed as $NaNO_3$
  (5) Residual amount of nitrite formed from nitrate included, expressed as $NaNO_2$
  (6) Propionic acid and its salls may be present in certain fermented products resutling from the fermentation process following good manufacturing practice

  4) The antioxidants listed in Table 4.2.15 may only be used as specified in the table and only up to the maximum level indicated.

Table 4.2.15

EC No. and name	Food	Maximum level (mg/kg)
E 310 Propyl gallate E 311 Octyl gallate E 312 Dodecyl gallate E 320 Butylated hydroxyanisole (BHA) E 321 Butylated hydroxytoluene (BHT)	Fats and oils for the professional manufacture of heat-treated foods Frying oil and frying fat, excluding olive pomace oil Lard; fish oil; beef, poultry and sheep fat	200 (1) (gallates and BHA, individually or in combination) and 100 (1) (BHT), both expressed on fat
	Cake mixes Cereal-based snack foods Milk products for vending machines Dehydrated soups and broths Sauces Dehydrated meat Processed nuts Seasonings and condiments Precooked cereals	200 (gallates and BHA, individually or in combination), expressed on fat
	Dehydrated granulated potatoes	25 (gallates and BHA, individually or in combination)
	Chewing gum Dietary supplements	400 (gallates, BHT and BHA, individually or in combination)
E 315 Erythorbic accid E 316 Sodium erythorbate	Semi-preserved and preserved meat products	500, expressed as erythorbic acid
	Preserved and semi-preserved fish products Frozen and deep-frozen fish with red skin	1500 expressed as erythorbic acid

Note: (1) When combinations of gallates, BHA and BHT are used, the individual levels must be reduced proportionally

Other permitted miscellaneous additives
The miscellaneous additives listed in Table 4.2.16 may only be used as specified in the table and only up to the maximum level indicated.

4.2

Table 4.2.16

EC No. and name	Food	Maximum level
E 297 Fumaric acid	(*pro memoria*) Wine in accordance with Regulation (EEC) No. 1873/84 authorising the offer or disposal for direct human consumption of certain imported wines which may have undergone oenological processes not provided for in Regulation (EEC) No. 337/79	
	Fillings and toppings for fine bakery wares	2.5 g/kg
	Sugar confectionery	1 g/kg
	Gel-like desserts Fruit-flavoured desserts Dry powdered dessert mixes	4 g/kg
	Instant powders for fruit-based drinks	1 g/l
	Instant tea powder	1 g/l
	Chewing gum	2 g/kg

In the following applications, the indicated maximum quantities of phosphoric acid and the phosphates E 338, E 339, E 340, E 341, E 450. E 451 and E 452 may be added individually or in combination (expressed as P_2O_5)

E 338 Phosphoric acid	Non-alcoholic flavoured drinks	700 mg/l (E 338 only)
E 339 Sodium phosphates:	Sterilised and UHT milk	1 g/l
(i) Monosodium phosphate	Partly dehydrated milk with less than 28% solids	1 g/kg
(ii) Disodium phosphate		
(iii) Trisodium phosphate	Partly dehydrated milk with more than 28% solids	1.5 g/kg
E 340 Potassium phosphates:	Dried milk and dried skimmed milk	2.5 g/kg
(i) Monopotassium phosphate		
(ii) Dipotassium phosphate		
(iii) Tripotassium phosphate	Pasteurised, sterilised and UHT creams	5 g/kg
E 341 Calcium phosphates:		
(i) Monocalcium phosphate	Whipped cream and vegetable fat analogues	5 g/kg
(ii) Dicalcium phosphate		
(iii) Tricalcium phosphate	Unripened cheese (except *mozzarella*)	2 g/kg
E 450 Diphosphates:		
(i) Disodium diphosphate	Processed cheese and processed cheese analogues	20 g/kg
(ii) Trisodium diphosphate		
(iii) Tetrasodium diphosphate	Meat products	5 g/kg
(iv) Dipotassium diphosphate	Sports drinks and prepared table waters	0.5 g/l
(v) Tetrapotassium diphosphate		
(vi) Dicalcium diphosphate	Dietary supplements	*quantum satis*
(vii) Calcium dihydrogen diphosphate	Salt and its substitutes	10 g/kg
	Vegetable protein drinks	20 g/l
E 451 Triphosphates:	Beverage whiteners	30 g/kg
(i) Pentasodium triphosphate	Beverage whiteners for vending machines	50 g/kg
(ii) Pentapotassium triphosphate	Edible ices	1 g/kg

4.2

EC No. and name	Food	Maximum level
E 452 Polyphosphates (i) Sodium polyphosphate (ii) Potassium polyphosphate (iii) Sodium calcium polyphosphate (iv) Calcium polyphosphates	Desserts	3 g/kg
	Dry powdered dessert mixes	7 g/kg
	Fine bakery wares	20 g/kg
	Flour	2.5 g/kg
	Flour, self raising	20 g/kg
	Soda bread	20 g/kg
	Liquid egg (white, yolk or whole egg)	10 g/kg
	Sauces	5 g/kg
	Soups and broths	3 g/kg
	Tea and herbal infusions	2 g/l
	Cider and perry	2 g/l
	Chewing gum	*quantum satis* (E 341 (ii) only)
	Dried powdered foods	10 g/kg (E 341 (iii) only)
	Chocolate and malt dairy-based drinks	2 g/l
	Alcoholic drinks (excluding wine and beer)	1 g/l
	Breakfast cereals	5 g/kg
	Snacks	5 g/kg
	Surimi	1 g/kg
	Fish and crustacean paste	5 g/kg
	Toppings (syrups for pancakes flavoured syrups for milkshakes and ice cream; similar products)	3 g/kg
	Special formulae for particular nutritional uses	5 g/kg
	Glazings for meat and vegetable products	4 g/kg
	Sugar confectionery	5 g/kg
	Icing sugar	10 g/kg
	Noodles	2 g/kg
	Batters	5 g/kg
	Fillets of unprocessed fish, frozen and deep-frozen	5 g/kg
	Frozen and deep-frozen crustacean products	5 g/kg
	Processed potato products (including frozen, deep-frozen, chilled and dried processed products)	5 g/kg
E 431 Polyoxyethylene (40) stearate	(*pro memoria*) Wine in accordance with Regulation (EEC) No. 1873/84 authorising the offer or disposal for direct human consumption of certain imported wines which may have undergone oenological processes not provided for in Regulation (EEC) No. 337/79	

4.2 *(continued)*

EC No. and name	Food	Maximum level
E 353 Metatartaric acid	Wine in accordance with Regulations (EEC) No. 822/87, (EEC) No. 4252/88, (EEC) No. 2332/92 and (EEC) No. 1873/84 and their implementing regulations	
	Made wine	100 mg/l
E 355 Adipic acid	Fillings and toppings for fine bakery wares	2 g/kg
E 356 Sodium adipate	Dry powdered dessert mixes	1 g/kg
E 357 Potassium adipate	Gel-like desserts	6 g/kg
	Fruit-flavoured desserts	1 g/kg
	Powders for home preparation of drinks	10 g/l expressed as adipic acid)
E 363 Succinic acid	Desserts	6 g/kg
	Soups and broths	5 g/kg
	Powders for home preparation of drinks	3 g/l
E 385 Calcium disodium ethylene diamine tetra-acetate (calcium disodium EDTA)	Emulsified sauces	75 mg/kg
	Canned and bottled pulses, legumes, mushrooms and artichokes	250 mg/kg
	Canned and bottled crustaceans and molluscs	75 mg/kg
	Canned and bottled fish	75 mg/kg
	Minarine	100 mg/kg
	Frozen and deep-frozen crustaceans	75 mg/kg
E 405 Propane-1,2-diol alginate	Fat emulsions	3 g/kg
	Fine bakery wares	2 g/kg
	Fillings, toppings and coatings for fine bakery wares and desserts	5 g/kg
	Sugar confectionery	1.5 g/kg
	Water-based edible ices	3 g/kg
	Cereal- and potato-based snacks	3 g/kg
	Sauces	8 g/kg
	Beer	100 mg/l
	Chewing gum	5 g/kg
	Fruit and vegetable preparations	5 g/kg
	Non-alcoholic flavoured drinks	300 mg/l
	Emulsified liqueur	10 g/l
	Dietetic foods intended for special medical purposes – dietetic formulae for weight control intended to replace total daily food intake or an individual meal	1.2 g/kg
	Dietary food supplements	1 g/kg

4.2

(continued)

EC No. and name	Food	Maximum level
E 416 Karaya gum	Cereal- and potato-based snacks	5 g/kg
	Nut coatings	10 g/kg
	Fillings, toppings and coatings for fine bakery wares	5 g/kg
	Desserts	6 g/kg
	Emulsified sauces	10 g/kg
	Egg-based liqueurs	10 g/l
	Dietary food supplements	quantum satis
	Chewing gum	5 g/kg

E 420   Sorbitol:
    (i) Sorbitol
    (ii) Sorbitol syrup
	Foods in general (except drinks and those foods referred to in Tables 4.2.18, 4.2.19 and in follow-on formulae and infant foods	

E 421 Mannitol	Frozen and deep-frozen	quantum satis
E 953 Isomalt	unprocessed fish, crustaceans,	(for purposes

E 965   Maltitol:
    (i) Maltitol
    (ii) Maltitol syrup
	molluscs and cephalopods
	Liqueurs
		other than
		sweetening)

E 966 Lactitol		
E 967 Xylitol		
E 432 Polyoxyethylene sorbitan monolaurate (polysorbate 20)	Fine bakery wares	3 g/kg
	Fat emulsions for baking purposes	10 g/kg
E 433 Polyoxyethylene sorbitan monooleate (polysorbate 80)	Milk and cream analogues	5 g/kg
	Edible ices	1 g/kg
E 434 Polyoxyethylene sorbitan monopalmitate (polysorbate 40)	Desserts	3 g/kg
	Sugar confectionery	1 g/kg
E 435 Polyoxyethylene sorbitan monostearate (polysorbate 60)	Emulsified sauces	5 g/kg
	Soups	1 g/kg
E 436 Polyoxyethylene sorbitan tristearate (polysorbate 65)	Chewing gum	5 g/kg
	Dietary food supplements	quantum satis
	Dietetic foods intended for special medical purposes – dietetic formulae for weight control intended to replace total daily food intake or an individual meal	1 g/kg Individually or in combination
E 442 Ammonium phosphatides	Cocoa and chocolate products as defined in Directive 73/241	10 g/kg
	Cocoa-based confectionery	10 g/kg
E 444 Sucrose acetate isobutyrate	Non-alcoholic flavoured cloudy drinks	300 mg/l
E 445 Glycerol esters of wood rosins	Non-alcoholic flavoured cloudy drinks	100 mg/l
E 473 Sucrose esters of fatty acids	Canned liquid coffee	1 g/l
E 474 Sucroglycerides	Heat-treated meat products	5 g/kg (on fat)
	Fat emulsions for baking purposes	10 g/kg
	Fine bakery wares	10 g/kg
	Beverage whiteners	20 g/kg

4.2 *(continued)*

EC No. and name	Food	Maximum level
	Edible ices	5 g/kg
	Sugar confectionery	5 g/kg
	Desserts	5 g/kg
	Sauces	10 g/kg
	Soups and broths	2 g/kg
	Fresh fruits, surface treatment	*quantum satis*
	Non-alcoholic aniseed-based drinks	5 g/l
	Non-alcoholic coconut and almond drinks	5 g/l
	Spirituous beverages (excluding wine and beer)	5 g/l
	Powders for the preparation of hot beverages	10 g/l
	Dairy-based drinks	5 g/l
	Dietary food supplements	*quantum satis*
	Dietetic foods intended for special medical purposes – Dietetic formulae for weight control intended to replace total daily food intake or an individual meal	5 g/kg
	Chewing gum	10 g/kg
		Individually or in combination
E 475 Polyglycerol esters of fatty acids	Fine bakery wares	10 g/kg
	Emulsified liqueurs	5 g/l
	Egg products	1 g/kg
	Beverage whiteners	0.5 g/kg
	Chewing gum	5 g/kg
	Fat emulsions	5 g/kg
	Milk and cream analogues	5 g/kg
	Sugar confectionery	2 g/kg
	Desserts	2 g/kg
	Dietary food supplements	*quantum satis*
	Dietetic foods intended for special medical purposes – dietetic formulae for weight control intended to replace total daily food intake or an individual meal	5 g/kg
	Granola-type breakfast cereals	10 g/kg
E 476 Polyglycerol polyricinoleate	Low and very low fat spreads and dressings	4 g/kg
	Cocoa-based confectionery, including chocolate	5 g/kg
E 477 Propane-1,2-diol esters of fatty acids	Fine bakery wares	5 g/kg
	Fat emulsions for baking purposes	10 g/kg
	Milk and cream analogues	5 g/kg

(*continued*)

EC No. and name	Food	Maximum level
	Beverage whiteners	1 g/kg
	Edible ices	3 g/kg
	Sugar confectionery	5 g/kg
	Desserts	5 g/kg
	Whipped dessert toppings other than cream	30 g/kg
	Dietetic foods intended for special medical purposes – dietetic formulae for weight control intended to replace total daily food intake or an individual meal	1 g/kg
E 479b Thermally oxidised soya bean oil interacted with mono- and diglycerides of fatty acids	Fat emulsions for frying purposes	5 g/kg
E 481 Sodium stearoyl-2-lactylate	Fine bakery wares	5 g/kg
E 482 Calcium stearoyl-2-lactylate	Quick-cook rice	4 g/kg
	Breakfast cereals	5 g/kg
	Emulsified liqueur	8 g/l
	Spirits with less than 15% alcohol by volume	8 g/l
	Cereal-based snacks	2 g/kg
	Chewing gum	2 g/kg
	Fat emulsions	10 g/kg
	Desserts	5 g/kg
	Sugar confectionery	5 g/kg
	Beverage whiteners	3 g/kg
	Cereal- and potato-based snacks	5 g/kg
	Minced and diced canned meat products	4 g/kg
	Powders for the preparation of hot beverages	2 g/l
	Dietetic foods intended for special medical purposes – Dietetic formulae for weight control intended to replace total daily food intake or an individual meal	2 g/kg
	Bread (except that referred to in Table 4.2.19)	3 g/kg
	Mostarda di frutta	2 g/kg
		Individually or in combination
E 483 Stearyl tartrate	Bakery wares (except breads referred to in Table 4.2.19)	4 g/kg
	Desserts	5 g/kg
E 491 Sorbitan monostearate ⎫	Fine bakery wares	10 g/kg
E 492 Sorbitan tristearate ⎬	Toppings and coatings for fine bakery wares	5 g/kg
E 493 Sorbitan monolaurate ⎭		

4.2

(continued)

EC No. and name	Food	Maximum level
E 494 Sorbitan monooleate ⎫ E 495 Sorbitan monopalmitate ⎭	Jelly marmalade	25 mg/kg (E 493 only)
	Fat emulsions	10 g/kg
	Milk and cream analogues	5 g/kg
	Beverage whiteners	5 g/kg
	Liquid tea concentrates and liquid fruit and herbal infusion concentrates	0 5 g/l
	Edible ices	0 5 g/kg
	Desserts	5 g/kg
	Sugar confectionery	5 g/kg
	Cocoa-based confectionery, including chocolate	10 g/kg (E 492 only)
	Emulsified sauces	5 g/kg
	Dietary food supplements	quantum satis
	Yeast for baking	quantum satis
	Chewing gum	5 g/kg
	Dietetic foods intended for special medical purposes – dietetic formulae for weight control intended to replace total daily food intake or an individual meal	5 g/kg Individually or in combination
	(pro memoria) For E 491 only, wine in accordance with Regulation (EEC) No 1873/84 authorising the offer or disposal for direct human consumption of certain imported wines which may have undergone oenological combination processes not provided for in Regulation (EEC) No 337/79	
E 512 Stannous chloride	Canned and bottled white asparagus	25 mg/kg (as tin)
E 520 Aluminium sulphate	Egg white	30 mg/kg
E 521 Aluminium sodium sulphate	Candied, crystallised and glacé fruit and vegetables	200 mg/kg
E 522 Aluminium potassium sulphate		
E 523 Aluminium ammonium sulphate		Individually or in combination, expressed as aluminium
E 541 Sodium aluminium phosphate, acidic	Fine bakery wares (scones and sponge wares only)	1 g/kg (as aluminium)
E 535 Sodium ferrocyanide	Salt and its substitutes	20 mg/kg
E 536 Potassium ferrocyanide E 538 Calcium ferrocyanide		Individually or in combination, expressed as anhydrous potassium ferrocyanide

4.2

(continued)

EC No. and name		Food	Maximum level
E 551	Silicon dioxide	Dried powdered foods including sugars)	10 g/kg
E 552	Calcium silicate		
E 553a	(i) Magnesium silicate	Salt and its substitutes	10 g/kg
	(ii) Magnesium trisilicate (asbestos free)	Dietary food supplements	*quantum satis*
E 553b	Talc (asbestos free)	Foods in tablet and coated tablet form	*quantum satis*
E 554	Sodium aluminium silicate		
E 555	Potassium aluminium silicate	Sliced hard cheese and sliced processed cheese	10 g/kg
E 556	Calcium aluminium silicate		Individually or in combination
E 559	Aluminium silicate (kaolin)		
		Chewing gum	
		Rice	
		Sausages (surface treatment only)	*quantum satis* (E 553b only)
		Moulded jelly sweets (surface treatment only)	
E 579	Ferrous gluconate	Olives darkened by oxidation	150 mg/kg (as iron)
E 585	Ferrous lactate		
E 620	Glutamic acid	Foods in general (except those referred to in Tables 4.2.18, 4.2.19 and in the follow-on formulae and infant foods	10 g/kg Individually or in combination
E 621	Monosodium glutamate		
E 622	Monopotassium glutamate		
E 623	Calcium diglutamate		
E 624	Monoammonium glutamate	Condiments and seasonings	*quantum satis*
E 625	Magnesium diglutamate		
E 626	Guanylic acid	Foods in general (except those referred to in Tables 4.2.18 4.2.19 and in the follow-on formulae and infant foods	500 mg/kg Individually or in combination, expressed as guanylic acid
E 627	Disodium guanylate		
E 628	Dipotassium guanylate		
E 629	Calcium guanylate		
E 630	Inosinic acid		
E 631	Disodium inosinate	Seasonings and condiments	*quantum satis*
E 632	Dipotassium inosinate		
E 633	Calcium inosinate		
E 634	Calcium 5'-ribonucleotides		
E 635	Disodium 5'-ribonucleotides		
E 900	Dimethyl polysiloxane	Jam, jellies and marmalades as defined in Directive 79/693/EEC and similar fruit spreads, including low calorie products	10 mg/kg
		Soups and broths	10 mg/kg
		Oils and fats for frying	10 mg/kg
		Confectionery (excluding chocolate)	10 mg/kg
		Non-alcoholic flavoured drinks	10 mg/l
		Pineapple juice	10 mg/l
		Canned and bottled fruit and vegetables	10 mg/kg
		Chewing gum	100 mg/kg
		(pro memoria) Wine in accordance with Regulation (EEC) No. 1873/ 84 authorising the offer or disposal for direct human consumption of	

4.2

(continued)

EC No. and name	Food	Maximum level
	certain imported wines which may have undergone oenological processes not provided for in Regulation (EEC) No. 337/79	
	søD ... *saft*	10 mg/l
	Batters	10 mg/kg
E 901 Beeswax, white and yellow E 902 Candelilla wax E 903 Carnauba wax E 904 Shellac	As glazing agents only for: – Confectionery (including chocolate) – Small products of fine bakery wares coated with chocolate – Snacks – Nuts – Coffee beans	*quantum satis*
	Dietary food supplements	*quantum satis*
	Fresh citrus fruits, melons, apples and pears (surface treatment only)	*quantum satis*
E 912 Montan acid esters E 914 Oxidised polyethylene wax	Fresh citrus fruits (surface treatment only)	*quantum satis*
E 927b Carbamide	Chewing gum without added sugars	30 g/kg
E 950 Acesulfame-K E 951 Aspartame E 957 Thaumatin	Chewing gum with added sugars	800 mg/kg (1) 2500 mg/kg (1) 10 mg/kg (1) (as flavour enhancer only)
E 959 Neohesperidine DC	Chewing gum with added sugars	150 mg/kg (1)
	Margarine Minarine Meat products Fruit jellies Vegetable proteins	5 mg/kg (as flavour enhancer only)
E 999 Quillaia extract	Water-based flavoured non-alcoholic drinks	200 mg/l calculated as anhydrous extract
E 1201 Polyvinylpyrrolidone E 1202 Polyvinylpolypyrrolidone	Dietary food supplements in tablet and coated tablet form	*quantum satis*
E 1505 Triethyl citrate	Dried egg white	*quantum satis*
Propane (2) Butane (2) iso-Butane (2)	Garlic-flavoured oil spray for producing garlic bread and pizza Vegetable oil pan spray for professional use only	*quantum satis*

Notes:
(1) If E 950, E 951, E 957 and E 959 are used in combination in chewing gum, the maximum level for each is reduced proportionally
(2) Authorised until 31 December 1997 in accordance with Article 5 of Directive 89/107/EEC pending consideration for inclusion in Directive 95/2/EC

Permitted carriers and solvents

Only the carriers and carrier solvents listed in Table 4.2.17 may be used, and where specified in the table, subject to the specified restrictions.

4.2

Table 4.2.17

EC No. and name		*Restricted use*
	Propan-1,2-diol (propylene glycol)	Colours, emulsifiers, antioxidants and enzymes (maximum 1 g/kg in or on the food)
E 422	Glycerol	
E 420	Sorbitol	
E 421	Mannitol	
E 953	Isomalt	
E 965	Maltitol	
E 966	Lactitol	
E 967	Xylitol	
E 400-404	Alginic acid and its sodium, potassium, calcium and ammonium salts	
E 405	Propan-1,2-diol alginate	
E 406	Agar	
E 407	Carrageenan	
E 410	Locust bean gum	
E 412	Guar gum	
E 413	Tragacanth	
E 414	Acacia gum (gum arabic)	
E 415	Xanthan gum	
E 440	Pectins	
E 432	Polyoxyethylene sorbitan monolaurate (polysorbate 20)	
E 433	Polyoxyethylene sorbitan monooleate (polysorbate 80)	
E 434	Polyoxyethylene sorbitan monopalmitate (polysorbate 40)	Antifoaming agents, colours and fat-soluble antioxidants
E 435	Polyoxyethylene sorbitan monostearate (polysorbate 60)	
E 436	Polyoxyethylene sorbitan tristearate (polysorbate 65)	
E 442	Ammonium phosphatides	Antioxidants
E 460	Cellulose (microcrystalline or powdered)	
E 461	Methyl cellulose	
E 463	Hydroxypropyl cellulose	
E 464	Hydroxypropyl methyl cellulose	
E 465	Ethyl methyl cellulose	
E 466	Carboxy methyl cellulose Sodium carboxy methyl cellulose	
E 322	Lecithins	
E 470b	Magnesium salts of fatty acids	
E 471	Mono- and diglycerides of fatty acids	
E 472a	Acetic acid esters of mono- and diglycerides of fatty acids	

113

4.2

(*continued*)

EC No. and name		*Restricted use*
E 472c	Citric acid esters of mono- and diglycerides of fatty acids	Colours and fat-soluble antioxidants
E 472e	Mono- and diacetyl tartaric acid esters of mono- and diglycerides of fatty acids	
E 473	Sucrose esters of fatty acids	
E 475	Polyglycerol esters of fatty acids	
E 491	Sorbitan monostearate	
E 492	Sorbitan tristearate	
E 493	Sorbitan monolaurate	Colours and anti-foaming agents
E 494	Sorbitan monooleate	
E 495	Sorbitan monopalmitate	
E 1404	Oxidised starch	
E 1410	Monostarch phosphate	
E 1412	Distarch phosphate	
E 1413	Phosphated distarch phosphate	
E 1414	Acetylated distarch phosphate	
E 1420	Acetylated starch	
E 1422	Acetylated distarch adipate	
E 1440	Hydroxypropyl starch	
E 1442	Hydroxypropyl distarch phosphate	
E 1450	Starch sodium octenyl succinate	
E 170	Calcium carbonates	
E 263	Calcium acetate	
E 331	Sodium citrates	
E 332	Potassium citrates	
E 341	Calcium phosphates	
E 501	Potassium carbonates	
E 504	Magnesium carbonates	
E 508	Potassium chloride	
E 509	Calcium chloride	
E 511	Magnesium chloride	
E 514	Sodium sulphate	
E 515	Potassium sulphate	
E 516	Calcium sulphate	
E 517	Ammonium sulphate	
E 577	Potassium gluconate	
E 640	Glycine and its sodium salt	
E 1505	Triethyl citrate	
E 1518	Glyceryl triacetate (triacetin)	
E 551	Silicon dioxide	Emulsifiers and colours, max. 5%
E 552	Calcium silicate	
E 553b	Talc	
E 558	Bentonite	Colours, max. 5%
E 559	Aluminium silicate (kaolin)	
E 901	Beeswax	Colours
E 1200	Polydextrose	
E 1201	Polyvinylpyrrolidone	Sweeteners
E 1202	Polyvinylpolypyrrolidone	

Food in which the miscellaneous additives in Table 4.2.8 are generally prohibited
The foods listed in Table 4.2.18 are not permitted to contain the
miscellaneous additives listed in Table 4.2.8.

Table 4.2.18

Unprocessed foods
Honey
Non-emulsified oils and fats of animal or vegetable origin
Butter
Pasteurised and sterilised (including UHT sterilisation) milk and cream (including
  skimmed, plain and semi-skimmed)
Unflavoured, live fermented milk products
Natural mineral water and spring water
Coffee (including flavoured instant coffee) and coffee extracts
Unflavoured leaf tea
Sugars)
Dry pasta
Natural unflavoured buttermilk (excluding sterilised buttermilk)

---

*Food in which a limited number of the miscellaneous additives in Table 4.2.8 may
be used*
The foods listed in Table 4.2.19 may only use the miscellaneous additives
listed in Table 4.2.8 when indicated in the table.

Table 4.2.19

Food	Additive		Maximum level
Cocoa and chocolate products as defined in Directive 73/241 (except cocoa and chocolate products energy-reduced or with no added sugars)	E 330	Citric acid	0.5%
	E 322	Lecithins	*quantum satis*
	E 334	Tartaric acid	0 5%
	E 422	Glycerol	
	E 471	Mono- and diglycerides of fatty acids	*quantum satis*
	E 170	Calcium carbonates	
	E 500	Sodium carbonates	
	E 501	Potassium carbonates	
	E 503	Ammonium carbonates	7% on dry matter
	E 504	Magnesium carbonates	without fat,
	E 524	Sodium hydroxide	expressed as
	E 525	Potassium hydroxide	potassium
	E 526	Calcium hydroxide	carbonates
	E 527	Ammonium hydroxide	
	E 528	Magnesium hydroxide	
	E 530	Magnesium oxide	
	E 414	Acacia gum	as glazing agents
	E 440	Pectins	only, *quantum satis*
Fruit juices and nectars as defined in Directive 93/77	E 300	Ascorbic acid	*quantum satis*

4.2

(continued)

Food	Additive	Maximum level
Pineapple juice as defined in Directive 93/77	E 296 Malic acid	3 g/l
Nectars as defined in Directive 93/77	E 330 Citric acid E 270 Lactic acid	5 g/l 5 g/l
Grape juice as defined in Directive 93/77	E 170 Calcium carbonates E 336 Potassium tartrates	} *quantum satis*
Fruit juices as defined in Directive 93/77	E 330 Citric acid	3 g/l
Extra jam and extra jelly as defined in Directive 79/693	E 270 Lactic acid E 296 Malic acid E 300 Ascorbic acid E 327 Calcium lactate E 330 Citric acid E 331 Sodium citrates E 333 Calcium citrates E 334 Tartaric acid E 335 Sodium tartrates E 350 Sodium malates E 440 Pectins E 471 Mono- and diglycerides of fatty acids	} *quantum satis*
Jam, jellies and marmalades as defined in Directive 79/693 and other similar fruit spreads including low-calorie products	E 270 Lactic acid E 296 Malic acid E 300 Ascorbic acid E 327 Calcium lactate E 330 Citric acid E 331 Sodium citrates E 333 Calcium citrates E 334 Tartaric acid E 335 Sodium tartrates E 350 Sodium malates	} *quantum satis*
	E 400 Alginic acid E 401 Sodium alginate E 402 Potassium alginate E 403 Ammonium alginate E 404 Calcium alginate E 406 Agar E 407 Carrageenan E 410 Locust bean gum E 412 Guar gum E 415 Xanthan gum E 418 Gellan gum	} 10 g/kg (individually of in combination)
	E 440 Pectins E 509 Calcium chloride E 524 Sodium hydroxide	} *quantum satis*

(continued)

Food	Additive	Maximum level
Partially dehydrated and dehydrated milk as defined in Directive 76/118	E 300 Ascorbic acid E 301 Sodium ascorbate E 304 Fatty acid esters of ascorbic acid E 322 Lecithins E 331 Sodium citrates E 332 Potassium citrates E 407 Carrageenan E 500 (ii) Sodium bicarbonate E 501 (ii) Potassium bicarbonate E 509 Calcium chloride	*quantum satis*
Sterilised, pasteurised and UHT cream, low-calorie cream and pasteurised low-fat cream	E 270 Lactic acid E 322 Lecithins E 325 Sodium lactate E 326 Potassium lactate E 327 Calcium lactate E 330 Citric acid E 331 Sodium citrates E 332 Potassium citrates E 333 Calcium citrates E 400 Alginic acid E 401 Sodium alginate E 402 Potassium alginate E 403 Ammonium alginate E 404 Calcium alginate E 406 Agar E 407 Carrageenan E 410 Locust bean gum E 415 Xanthan gum E 440 Pectins E 460 Celluloses E 461 Methyl cellulose E 463 Hydroxypropyl cellulose E 464 Hydroxypropyl methyl cellulose E 465 Ethyl methyl cellulose E 466 Carboxy methyl cellulose Sodium carboxy methyl cellulose E 471 Mono- and diglycerides of fatty acids E 508 Potassium chloride E 509 Calcium chloride E 1404 Oxidised starch E 1410 Monostarch phosphate E 1412 Distarch phosphate E 1413 Phosphated distarch phosphate E 1414 Acetylated distarch phosphate E 1420 Acetylated starch E 1422 Acetylated distarch adipate E 1440 Hydroxypropyl starch E 1442 Hydroxypropyl distarch phosphate E 1450 Starch sodium octenyl succinate	*quantum satis*

4.2

(*continued*)

Food	Additive		Maximum level
Frozen and deep-frozen unprocessed fruit and vegetables Fruit compote Unprocessed fish, crustaceans and molluscs, including such products frozen and deep-frozen	E 300 E 301 E 302 E 330 E 331 E 332 E 333	Ascorbic acid Sodium ascorbate Calcium ascorbate Citric acid Sodium citrates Potassium citrates Calcium citrates	*quantum satis*
Quick-cook rice	E 471 E 472a	Mono- and diglycerides of fatty acids Acetic acid esters of mono- and diglycerides of fatty acids	*quantum satis*
Non-emulsified oils and fats of animal or vegetable origin (except virgin oils and olive oils)	E 304 E 306 E 307 E 308 E 309 E 322 E 471 E 330 E 331 E 332 E 333	Fatty acid esters of ascorbic acid Tocopherol-rich extract α-Tocopherol γ-Tocopherol δ-Tocopherol Lecithins Mono- and diglycerides of fatty acids Citric acid Sodium citrates Potassium citrates Calcium citrates	*quantum satis* 30 g/l 10 g/l *quantum satis*
Refined olive oil, including olive pomace oil	E 307	α-Tocopherol	200 mg/l
Ripened cheese	E 170 E 504 E 509 E 575	Calcium carbonates Magnesium carbonates Calcium chloride Glucono-δ-lactone	*quantum satis*
Mozzarella and whey cheese	E 270 E 330 E 575	Lactic acid Citric acid Glucono-δ-lactone	*quantum satis*
Canned and bottled fruit and vegetables	E 260 E 261 E 262 E 263 E 270 E 300 E 301 E 302 E 325 E 326 E 327 E 330 E 331 E 332 E 333 E 334	Acetic acid Potassium acetate Sodium acetates Calcium acetate Lactic acid Ascorbic acid Sodium ascorbate Calcium ascorbate Sodium lactate Potassium lactate Calcium lactate Citric acid Sodium citrates Potassium citrates Calcium citrates Tartaric acid	*quantum satis*

(*continued*)

Food	Additive	Maximum level
	E 335 Sodium tartrates	
	E 336 Potassium tartrates	
	E 337 Sodium potassium tartrate	
	E 509 Calcium chloride	
	E 575 Glucono-δ-lactone	
Gehakt	E 330 Citric acid	
	E 331 Sodium citrates	*quantum satis*
	E 332 Potassium citrates	
	E 333 Calcium citrates	
Prepacked preparations of fresh minced meat	E 300 Ascorbic acid	
	E 301 Sodium ascorbate	
	E 302 Calcium ascorbate	
	E 330 Citric acid	*quantum satis*
	E 331 Sodium citrates	
	E 332 Potassium citrates	
	E 333 Calcium citrates	
Bread prepared solely with the following ingredients: wheat flour, water, yeast or leaven, salt	E 260 Acetic acid	
	E 261 Potassium acetate	
	E 262 Sodium acetates	
	E 263 Calcium acetate	
	E 270 Lactic acid	
	E 300 Ascorbic acid	
	E 301 Sodium ascorbate	
	E 302 Calcium ascorbate	
	E 304 Fatty acid esters of ascorbic acid	
	E 322 Lecithins	
	E 325 Sodium lactate	
	E 326 Potassium lactate	
	E 327 Calcium lactate	*quantum satis*
	E 471 Mono- and diglycerides of fatty acids	
	E 472a Acetic acid esters of mono- and diglycerides of fatty acids	
	E 472d Tartaric acid esters of mono- and diglycerides of fatty acids	
	E 472e Mono- and diacetyl tartaric acid esters of mono- and diglycerides of fatty acids	
	E 472f Mixed acetic and tartaric acid esters of mono- and diglycerides of fatty acids	
Pain courant français	E 260 Acetic acid	
	E 261 Potassium acetate	
	E 262 Sodium acetates	
	E 263 Calcium acetate	
	E 270 Lactic acid	
	E 300 Ascorbic acid	
	E 301 Sodium ascorbate	
	E 302 Calcium ascorbate	
	E 304 Fatty acid esters of ascorbic acid	*quantum satis*

119

4.2

(continued)

Food	Additive	Maximum level
	E 322 Lecithins	
	E 325 Sodium lactate	
	E 326 Potassium lactate	
	E 327 Calcium lactate	
	E 471 Mono- and diglycerides of fatty acids	
Fresh pasta	E 270 Lactic acid	
	E 300 Ascorbic acid	
	E 301 Sodium ascorbate	
	E 322 Lecithins	
	E 330 Citric acid	*quantum satis*
	E 334 Tartaric acid	
	E 471 Mono- and diglycerides of fatty acids	
	E 575 Glucono-δ-lactone	
Wines and sparkling wines and partially fermented grape must	Additives authorised: in accordance with Regulations (EEC) No. 822/87, (EEC) No. 4252/88, (EEC) No. 2332/92 and (EEC) No. 1873/84 and their implementing regulations; in accordance with Regulation (EEC). No 1873/84 authorising the offer or disposal for direct human consumption of certain imported wines which may have undergone oenological processes not provided for in Regulation (EEC) No. 337/79	*pro memoria*
Beer	E 270 Lactic acid	
	E 300 Ascorbic acid	
	E 301 Sodium ascorbate	*quantum satis*
	E 330 Citric acid	
	E 414 Acacia gum	
Foie gras, foie gras entier, blocs de foie gras	E 300 Ascorbic acid	*quantum satis*
	E 301 Sodium ascorbate	

Miscellaneous additives permitted in foods for infants and young children

1) In addition to the additives listed in points (2)–(4), formulae and weaning foods for infants and young children may contain E 414 acacia gum (gum arabic) and E 551 silicon dioxide resulting from the addition of nutrient preparations containing not more than 10 g/kg of each of these substances, as well as E 421 mannitol when used as a carrier for vitamin B_{12} (not less than 1 part vitamin B_{12} to 1000 parts mannitol).

2) Only the miscellaneous additives listed in Table 4.2.20 may be used in infant formulae for infants (except that, for the manufacture of acidified milks, non-pathogenic L(+)-lactic acid producing cultures may be used):

Table 4.2.20

4.2

EC No. and name	Maximum level
E 270 Lactic acid (L(+)-form only)	*quantum satis*
E 330 Citric acid	*quantum satis*
E 338 Phosphoric acid	In conformity with the limits set in Annex I to Directive 91/321/EEC
E 306 Tocopherol-rich extract	
E 307 α-Tocopherol	10 mg/l individually or
E 308 γ-Tocopherol	in combination
E 309 δ-Tocopherol	
E 322 Lecithins	1 g/l (1)
E 471 Mono- and diglycerides of fatty acids	4 g/l (1)

Note: (1) If more than one of the substances E 322 and E 471 is added to a food, the maximum level established for that food for each of those substances is lowered with that relative part as is present of the other substance in that food

3) Only the miscellaneous additives listed in Table 4.2.21 may be used in follow-on formulae for infants (except that, for the manufacture of acidified milks, non-pathogenic L(+)-lactic acid producing cultures may be used):

Table 4.2.21

EC No. and name	Maximum level
E 270 Lactic acid (L(+)-form only)	*quantum satis*
E 330 Citric acid	*quantum satis*
E 306 Tocopherol-rich extract	
E 307 α-Tocopherol	10 mg/l individually or
E 308 γ-Tocopherol	in combination
E 309 δ-Tocopherol	
E 338 Phosphoric acid	In conformity with the limits set in Annex II to Directive 91/321/EEC
E 440 Pectins	5 g/l in acidified follow-on formulae only
E 322 Lecithins	1 g/l (1)
E 471 Mono- and diglycerides of fatty acids	4 g/l (l)
E 407 Carrageenan	0.3 g/l (2)
E 410 Locust bean gum	1 g/l (2)
E 412 Guar gum	1 g/l (2)

Notes:
(1) If more than one of the substances E 322 and E 471 is added to a food, the maximum level established for that food for each of those substances is lowered with that relative part as is present of the other substance in that food

(2) If more than one of the substances E 407, E 410 and E 412 is added to a food,

4.2

the maximum level established for that food for each of those substances is lowered with that relative part as is present of the other substances together in that food

4) Only the miscellaneous additives listed in Table 4.2.22 may be used in weaning foods for infants and young children and subject to the specified maximum level.

Table 4.2.22

EC No. and name		Food	Maximum level
E 170	Calcium carbonates	Weaning foods	*quantum satis*
E 260	Acetic acid		(only for pH
E 261	Potassium acetate		adjustment)
E 262	Sodium acetates		
E 263	Calcium acetate		
E 270	Lactic acid (1)		
E 296	Malic acid (1)		
E 325	Sodium lactate (1)		
E 326	Potassium lactate (1)		
E 327	Calcium lactate (1)		
E 330	Citric acid		
E 331	Sodium citrates		
E 332	Potassium citrates		
E 333	Calcium citrates		
E 507	Hydrochloric acid		
E 524	Sodium hydroxide		
E 525	Potassium hydroxide		
E 526	Calcium hydroxide		
E 500	Sodium carbonates	Weaning foods	*quantum satis*
E 501	Potassium carbonates		(only as raising
E 503	Ammonium carbonates		agents)
E 300	L-Ascorbic acid	Fruit- and vegetable-based drinks, juices and baby foods	0.3 g/kg
E 301	Sodium L-ascorbate		
E 302	Calcium L-ascorbate		
		Fat-containing cereal-based foods including biscuits and rusks	0.2 g/kg, individually or in combination, expressed as ascorbic acid
E 304	L-Ascorbyl palmitate	Fat-containing cereals, biscuits, rusks and baby foods	0.1 g/kg, individually or in combination
E 306	Tocopherol-rich extract		
E 307	α-Tocopherol		
E 308	γ-Tocopherol		
E 309	δ-Tocopherol		
E 338	Phosphoric acid	Weaning foods	1 g/kg as P_2O_5 (only for pH adjustment)
E 339	Sodium phosphates	Cereals	1 g/kg,

(*continued*)

4.2

EC No. and name		Food	Maximum level
E 340	Potassium phosphates		individually or in
E 341	Calcium phosphates		combination, expressed as P_2O_5
E 322	Lecithins	Biscuits and rusks Cereal-based foods Baby foods	} 10 g/kg
E 471	Mono- and diglycerides of fatty acids	Biscuits and rusks Cereal-based foods	} 5 g/kg, individually or in
E 472a	Acetic acid esters of mono- and diglycerides of fatty acids	Baby foods	combination
E 472b	Lactic acid esters of mono- and diglycerides of fatty acids		
E 472c	Citric acid esters of mono- and diglycerides of fatty acids		
E 400	Alginic acid	Desserts	} 0.5 g/kg,
E 401	Sodium alginate	Puddings	individually or in
E 402	Potassium alginate		combination
E 404	Calcium alginate		
E 410	Locust bean gum	Weaning foods	10 g/kg,
E 412	Guar gum		individually or in
E 414	Acacia gum (gum arabic)		combination
E 415	Xanthan gum	Gluten-free cereal-based	20 g/kg,
E 440	Pectins	foods	individually or in combination
E 551	Silicon dioxide	Dry cereals	2 g/kg
E 334	Tartaric acid (1)	Biscuits and rusks	5 g/kg
E 335	Sodium tartrate (1)		as a residue
E 336	Potassium tartrate (1)		
E 354	Calcium tartrate (1)		
E 450a	Disodium diphosphate		
E 575	Glucono-δ-lactone		
E 1404	Oxidised starch	Weaning foods	50 g/kg
E 1410	Monostarch phosphate		
E 1412	Distarch phosphate		
E 1413	Phosphated distarch phosphate		
E 1414	Acetylated distarch phosphate		
E 1420	Acetylated starch		
E 1422	Acetylated distarch adipate		
E 1450	Starch sodium octenyl succinate		

Note: (1)   L(+)-form only

4.2

Compound foods

1) Subject to (3) and (4), if a food which contains a permitted miscellaneous additive is used as an ingredient in a compound food in which the miscellaneous additive is not permitted, then the presence of the miscellaneous additive in the compound food is not an offence.

2) Subject to (4), foods may contain permitted miscellaneous additives outside the permitted amounts where the food is to be used solely in the preparation of a compound food and the resulting use in the compound food is permitted.

3) Item (1) does not apply to foods listed in Tables 4.2.18 and 4.2.19 above.

4) Items (1) and (2) do not apply to any food for infants or young children (referred to in Directive 89/398) except where specifically provided for.

Sweeteners

Regulation: Sweeteners in Food Regulations 1995 (1995/3123)
Amendment: Sweeteners in Food (Amendment) Regulations 1996 (1996/1477)

Definitions

Sweetener: any food additive which is used or intended to be used (a) to impart a sweet taste to food, or (b) as a table-top sweetener.

Sale and use of sweeteners

1) Only the sweeteners listed in Table 4.2.23 are permitted for use in or on any food.

Table 4.2.23

E 420	Sorbitol:
	(i) Sorbitol
	(ii) Sorbitol syrup
E 421	Mannitol
E 950	Acesulfame K
E 951	Aspartame
E 952	Cyclamic acid and its Na and Ca salts
E 953	Isomalt
E 954	Saccharin and its Na, K and Ca salts
E 957	Thaumatin
E 959	Neohesperidine DC
E 965	Maltitol:
	(i) Maltitol
	(ii) Maltitol syrup
E 966	Lactitol
E 967	Xylitol

2) The permitted sweeteners may only be used on specified foods up to specified maximum levels. The foods and the levels are as given in Table 4.2.24.

3) Permitted sweeteners may not be used in or on any food for infants

(children under the age of 12 months) or young children (children between 1 and 3 years).

4) No person shall sell any table-top sweetener unless it only contains permitted sweeteners specified in (1) and it is labelled with:

   i) 'X-based table-top sweetener' with X being substituted by the name of any permitted sweetener used;

   ii) 'excessive consumption may induce laxative effects' where it contains polyols;

   iii)'contains a source of phenlyalanine' where it contains aspartame.

Table 4.2.24

EC No.	Permitted sweetener (1,2)	Foods in or on which permitted sweeteners may be used	Maximum usable dose (3.4)
E 420	Sorbitol (i) Sorbitol (ii) Sorbitol syrup	*Desserts and similar products* – Water-based flavoured desserts, energy-reduced or with no added sugar	
E 421 E 953 E 965	Mannitol Isomalt Maltitol: (i) Maltitol (ii) Maltitol syrup	– Milk- and milk-derivative-based preparations, energy-reduced or with no added sugar – Fruit- and vegetable-based desserts, energy-reduced or with no added sugar	
E 966 E 967	Lactitol Xylitol	– Egg-based desserts, energy-reduced or with no added sugar	
		– Cereal-based desserts, energy-reduced or with no added sugar	
		– Breakfast cereals or cereal-based products, energy-reduced or with no added sugar	
		– Fat-based desserts, energy-reduced or with no added sugar	
		– Edible ices, energy-reduced or with no added sugar	*quantum satis*
		– Jams, jellies, marmalades and crystallized fruit, energy-reduced or with no added sugar	
		– Fruit preparations, energy-reduced or with no added sugar, with the exception of those intended for the manufacture of fruit-juice-based drinks	
		Confectionery – Confectionery with no added sugar	
		– Dried-fruit-based confectionery, energy-reduced or with no added sugar	
		– Starch-based confectionery, energy-reduced or with no added sugar	

4.2 *(continued)*

EC No.	Permitted sweetener (1,2)	Foods in or on which permitted sweeteners may be used	Maximum usable dose (3,4)
		– Chewing gum with no added sugar	
E 420		*Miscellaneous* – Cocoa-based products, energy-reduced or with no added sugar	
E 421		– Cocoa-, milk-, dried fruit- or fat- based sandwich spreads, energy-reduced or with no added sugar	
E 953			*quatum satis*
E 965		– Sauces	
E 966		– Mustard	
E 967 *(continued)*		– Fine bakery products, energy-reduced or with no added sugar	
		– Products intended for particular nutritional uses	
		– Solid food supplements/dietary integrators	
E 950	Acesulfame K	*Non-alcoholic drinks* – Water-based flavoured drinks, energy-reduced or with no added sugar	350 mg/l
		– Milk- and milk-derivative-based or fruit-juice-based drinks, energy-reduced or with no added sugar	350 mg/l
		Desserts and similar products – Water-based flavoured desserts, energy-reduced or with no added sugar	350 mg/kg
		– Milk- and milk-derivative-based preparations, energy-reduced or with no added sugar	350 mg/kg
		– Fruit- and vegetable-based desserts, energy-reduced or with no added sugar	350 mg/kg
		– Egg-based desserts, energy-reduced or with no added sugar	350 mg/kg
		– Cereal-based desserts, energy-reduced or with no added sugar	350 mg/kg
		– Fat-based desserts, energy-reduced or with no added sugar	350 mg/kg

(continued)

EC No.	Permitted sweetener (1,2)	Foods in or on which permitted sweeteners may be used	Maximum usable dose (3,4)
		Confectionery	
		– Confectionery with no added sugar	500 mg/kg
		– Cocoa- or dried-fruit-based confectionery, energy-reduced or with no added sugar	500 mg/kg
		– Starch-based confectionery, energy-reduced or with no added sugar	1000 mg/kg
		– Chewing gum with no added sugar	2000 mg/kg
		Miscellaneous	
		– 'Snacks': certain flavours of ready to eat, prepacked, dry, savoury starch products and coated nuts	350 mg/kg
		– Cocoa-, milk-, dried-fruit- or fat-based sandwich spreads, energy-reduced or with no added sugar	1000 mg/kg
		– Cider and perry	350 mg/l
		– Alcohol-free beer or with an alcohol content not exceeding 1.2% vol.	350 mg/l
		– 'Bière de table/tafelbier/table beer' (original wort content less than 6%) except for 'Obergäriges Einfachbier'	350 mg/l
		– Beers with a minimum acidity of 30 milli-equivalents expressed as NaOH	350 mg/l
		– Brown beers of the 'oud bruin' type	350 mg/l
		– Edible ices, energy-reduced or with no added sugar	800 mg/kg
		– Canned or bottled fruit, energy-reduced or with no added sugar	350 mg/kg
		– Energy-reduced jams, jellies and marmalades	1000 mg/kg
		– Energy-reduced fruit and vegetable preparations	350 mg/kg
		– Sweet–sour preserves of fruit and vegetables	200 mg/kg
		– Sweet–sour preserves and semi-preserves of fish and marinades of fish, crustaceans and molluscs	200 mg/kg
		– Sauces	350 mg/kg
		– Mustard	350 mg/kg
		– Fine bakery products for special nutritional uses	1000 mg/kg

4.2 *(continued)*

EC No.	Permitted sweetener (1,2)	Foods in or on which permitted sweeteners may be used	Maximum usable dose (3,4)
		– Complete formulae for weight control intended to replace total daily food intake or an individual meal	450 mg/kg
		– Complete formulae and nutritional supplements for use under medical supervision	450 mg/kg
		– Liquid food supplements/dietary integrators	350 mg/l
		– Solid food supplements/dietary integrators	500 mg/kg
		– Vitamins and dietary preparations	2000 mg/kg
E 951	Aspartame	*Non-alcoholic drinks*	
		– Water-based flavoured drinks, energy-reduced or with no added sugar	600 mg/l
		– Milk- and milk-derivative-based or fruit-juice-based drinks, energy-reduced or with no added sugar	600 mg/l
		Desserts and similar products	
		– Water-based flavoured desserts, energy-reduced or with no added sugar	1000 mg/kg
		– Milk- and milk-derivative-based preparations, energy-reduced or with no added sugar	1000 mg/kg
		– Fruit- and vegetable-based desserts energy-reduced or with no added sugar	1000 mg/kg
		– Egg-based desserts, energy-reduced or with no added sugar	1000 mg/kg
		– Cereal-based desserts, energy-reduced or with no added sugar	1000 mg/kg
		– Fat-based desserts, energy-reduced or with no added sugar	1000 mg/kg
		Confectionery	
		– Confectionery with no added sugar	1000 mg/kg
		– Cocoa- or dried-fruit-based confectionery, energy-reduced or with no added sugar	2000 mg/kg
		– Starch-based confectionery, energy-reduced or with no added sugar	2000 mg/kg
		– Chewing gum with no added sugar	5500 mg/kg

4.2

(continued)

EC No. Permitted sweetener (1,2)	Foods in or on which permitted sweeteners may be used	Maximum usable dose (3,4)
	Miscellaneous	
	– 'Snacks': certain flavours of ready to eat, prepacked, dry, savoury starch products and coated nuts	500 mg/kg
	– Cocoa-, milk-, dried-fruit- or fat-based sandwich spreads, energy-reduced or with no added sugar	1000 mg/kg
	– Cider and perry	600 mg/l
	– Alcohol-free beer or with an alcohol content not exceeding 1.2% vol.	600 mg/l
	– 'Bière de table/tafelbier/table beer' (original wort content less than 6%) except for 'Obergäriges Einfachbier'	600 mg/l
	– Beers with a minimum acidity of 30 milli-equivalents expressed as NaOH	600 mg/l
	– Brown beers of the 'oud bruin' type	600 mg/l
	– Edible ices, energy-reduced or with no added sugar	800 mg/kg
	– Canned or bottled fruit, energy-reduced or with no added sugar	1000 mg/kg
	– Energy-reduced jams, jellies and marmalades	1000 mg/kg
	– Energy-reduced fruit and vegetable preparations	1000 mg/kg
	– Sweet–sour preserves of fruit and vegetables	300 mg/kg
	– Sweet–sour preserves and semi-preserves of fish and marinades of fish, crustaceans and molluscs	300 mg/kg
	– Sauces	350 mg/kg
	– Mustard	350 mg/kg
	– Fine bakery products for special nutritional uses	1700 mg/kg
	– Complete formulae for weight control intended to replace total daily food intake or an individual meal	800 mg/kg
	– Complete formulae and nutritional supplements for use under medical supervision	1000 mg/kg
	– Liquid food supplements/dietary integrators	600 mg/kg

4.2

(continued)

EC No.	Permitted sweetener (1,2)	Foods in or on which permitted sweeteners may be used	Maximum usable dose (3,4)
		– Solid food supplements/dietary integrators	2000 mg/kg
		– Vitamins and dietary preparations	5500 mg/kg
E 952	Cyclamic acid and its Na and Ca salts (5)	*Non-alcoholic drinks* – Water-based flavoured drinks, energy-reduced or with no added sugar	400 mg/l
		– Milk- and milk-derivative-based or fruit-juice-based drinks, energy-reduced or with no added sugar	400 mg/l
		Desserts and similar products – Water-based flavoured desserts energy-reduced or with no added sugar	250 mg/kg
		– Milk- and milk-derivative-based preparations, energy-reduced or with no added sugar	250 mg/kg
		– Fruit- and vegetable-based desserts, energy-reduced or with no added sugar	250 mg/kg
		– Egg-based desserts, energy-reduced or with no added sugar	250 mg/kg
		– Cereal-based desserts, energy-reduced or with no added sugar	250 mg/kg
		– Fat-based desserts, energy-reduced or with no added sugar	250 mg/kg
		Confectionery – Confectionery with no added sugar	500 mg/kg
		– Cocoa- or dried-fruit-based confectionery, energy-reduced or with no added sugar	500 mg/kg
		– Starch-based confectionery, energy-reduced or with no added sugar	500 mg/kg
		– Chewing gum with no added sugar	1500 mg/kg
		Miscellaneous – Cocoa-, milk-, dried-fruit- or fat-based sandwich spreads, energy-reduced or with no added sugar	500 mg/kg
		– Edible ices, energy-reduced or with no added sugar	250 mg/kg
		– Canned or bottled fruit, energy-reduced or with no added sugar	1000 mg/kg

130

(continued)

4.2

EC No.	Permitted sweetener (1,2)	Foods in or on which permitted sweeteners may be used	Maximum usable dose (3,4)
		– Energy-reduced jams, jellies and marmalades	1000 mg/kg
		– Energy-reduced fruit and vegetable preparations	250 mg/kg
		– Fine bakery products for special nutritional uses	1600 mg/kg
		– Complete formulae for weight control intended to replace total daily food intake or an individual meal	400 mg/kg
		– Complete formulae and nutritional supplements for use under medical supervision	400 mg/kg
		– Liquid food supplements/dietary integrators	400 mg/kg
		– Solid food supplements/dietary integrators	500 mg/kg
E 954	Saccharin and its Na, K and Ca salts (6)	*Non-alcoholic drinks* – Water-based flavoured drinks, energy-reduced or with no added sugar	80 mg/l
		– Milk- and milk-derivative-based or fruit-juice-based drinks, energy-reduced or with no added sugar	80 mg/l
		– 'Gaseosa': non-alcoholic water-based drink with added carbon dioxide, sweeteners and flavourings	100 mg/l
		Desserts and similar products – Water-based flavoured desserts, energy-reduced or with no added sugar	100 mg/kg
		– Milk- and milk-derivative-based preparations, energy-reduced or with no added sugar	100 mg/kg
		– Fruit- and vegetable-based desserts, energy-reduced or with no added sugar	100 mg/kg
		– Egg-based desserts, energy-reduced or with no added sugar	100 mg/kg
		– Cereal-based desserts, energy-reduced or with no added sugar	100 mg/kg
		– Fat-based desserts, energy-reduced or with no added sugar	100 mg/kg

4.2 *(continued)*

EC No.	Permitted sweetener (1,2)	Foods in or on which permitted sweeteners may be used	Maximum usable dose (3,4)
		Confectionery	
		– Confectionery with no added sugar	500 mg/kg
		– Cocoa- or dried-fruit-based confectionery, energy-reduced or with no added sugar	500 mg/kg
		– Starch-based confectionery, energy-reduced or with no added sugar	300 mg/kg
		– Chewing gum with no added sugar	1200 mg/kg
		Miscellaneous	
		– 'Snacks': certain flavours of ready to eat, prepacked, dry, savoury starch products and coated nuts	100 mg/kg
		– Essoblaten	800 mg/kg
		– Cocoa-, milk-, dried-fruit- or fat-based sandwich spreads, energy-reduced or with no added sugar	200 mg/kg
		– Cider and perry	80 mg/l
		– Alcohol-free beer or with an alcohol content not exceeding 1.2% vol.	80 mg/l
		– 'Biere de table/tafelbier/table beer' (original wort content less than 6%) except for 'Obergäriges Einfachbier'	80 mg/l
		– Beers with a minimum acidity of 30 milli-equivalents expressed as NaOH	80 mg/l
		– Brown beers of the 'oud bruin' type	80 mg/l
		– Edible ices, energy-reduced or with no added sugar	100 mg/kg
		– Canned or bottled fruit, energy-reduced or with no added sugar	200 mg/kg
		– Energy-reduced jams, jellies and marmalades	200 mg/kg
		– Energy-reduced fruit and vegetable preparations	200 mg/kg
		– Sweet–sour preserves of fruit and vegetables	160 mg/kg
		– Sweet–sour preserves and semi-preserves of fish and marinades of fish, crustaceans and molluscs	160 mg/kg
		– Sauces	160 mg/kg
		– Mustard	320 mg/kg
		– Fine bakery products for special nutritional uses	170 mg/kg

(continued)

EC No.	Permitted sweetener (1,2)	Foods in or on which permitted sweeteners may be used	Maximum usable dose (3,4)
		– Complete formulae for weight control intended to replace total daily food intake or an individual meal	240 mg/kg
		– Complete formulae and nutritional supplements for use under medical supervision	200 mg/kg
		– Liquid food supplements/dietary integrators	80 mg/kg
		– Solid food supplements/dietary integrators	500 mg/kg
		– Vitamins and dietary preparations	1200 mg/kg
E 957	Thaumatin	*Confectionery*	
		– Confectionery with no added sugar	50 mg/kg
		– Cocoa- or dried-fruit-based confectionery, energy-reduced or with no added sugar	50 mg/kg
		– Chewing gum with no added sugar	50 mg/kg
		Miscellaneous	
		– Vitamins and dietary preparations	400 mg/kg
E 959	Neohesperidine DC	*Non-alcoholic drinks*	
		– Water-based flavoured drinks, energy-reduced or with no added sugar	30 mg/l
		– Milk- and milk-derivative-based drinks, energy-reduced or with no added sugar	50 mg/l
		– Fruit-juice-based drinks, energy-reduced or with no added sugar	30 mg/l
		Desserts and similar products	
		– Water-based flavoured desserts, energy-reduced or with no added sugar	50 mg/kg
		– Milk- and milk-derivative-based preparations, energy-reduced or with no added sugar	50 mg/kg
		– Fruit- and vegetable-based desserts, energy-reduced or with no added sugar	50 mg/kg
		– Egg-based desserts, energy-reduced or with no added sugar	50 mg/kg
		– Cereal-based desserts, energy-reduced or with no added sugar	50 mg/kg

4.2 *(continued)*

EC No. Permitted sweetener (1,2)	Foods in or on which permitted sweeteners may be used	Maximum usable dose (3,4)
	– Fat-based desserts, energy-reduced or with no added sugar	50 mg/kg
	Confectionery – Confectionery with no added sugar	100 mg/kg
	– Cocoa- or dried-fruit-based confectionery, energy-reduced or with no added sugar	100 mg/kg
	– Starch-based confectionery, energy-reduced or with no added sugar	150 mg/kg
	– Chewing gum with no added sugar	400 mg/kg
	Miscellaneous – Cocoa-, milk-, dried-fruit- or fat-based sandwich spreads, energy-reduced or with no added sugar	50 mg/kg
	– Cider and perry	20 mg/l
	– Alcohol-free beer or with an alcohol content not exceeding 1.2% vol.	10 mg/l
	– 'Bière de table/tafelbier/table beer' (original wort content less than 6%) except for 'Obergäriges Einfachbier'	10 mg/l
	– Beers with a minimum acidity of 30 milli-equivalents expressed as NaOH	10 mg/l
	– Brown beers of the 'oud bruin' type	10 mg/l
	– Edible ices, energy-reduced or with no added sugar	50 mg/kg
	– Canned or bottled fruit, energy-reduced or with no added sugar	50 mg/kg
	– Energy-reduced jams, jellies and marmalades	50 mg/kg
	– Sweet–sour preserves of fruit and vegetables	100 mg/kg
	– Energy-reduced fruit and vegetable preparations	50 mg/kg
	– Sweet–sour preserves and semi-preserves of fish and marinades of fish, crustaceans and molluscs	30 mg/kg
	– Sauces	50 mg/kg
	– Mustard	50 mg/kg
	– Fine bakery products for special nutritional uses	150 mg/kg

(*continued*)

4.2

EC No.	Permitted sweetener (1,2)	Foods in or on which permitted sweeteners may be used	Maximum usable dose (3,4)
		– Complete formulae for weight control intended to replace total daily food intake or an individual meal	100 mg/kg
		– Liquid food supplements/dietary integrators	50 mg/kg
		– Solid food supplements/dietary integrators	100 mg/kg

Notes:
  (1) The description 'with no added sugar' means that the food referred to does not contain any (i) added monosaccharide, (ii) added disaccharide, (iii) other added food used for its sweetening properties
  (2) The description 'energy-reduced' means that the food referred to has an energy value reduced by at least 30% compared with the original or a similar food
  (3) The maximum usable dose indicated is the maximum amount in mg sweetener/kg (or l) of food which is ready to eat having been prepared according to any instructions for use
  (4) The term *quantum satis* means that no maximum level of permitted sweetener is specified but that it is used in accordance with good manufacturing practice at a level not higher than is necessary to achieve the intended purpose and provided that such use does not mislead the consumer
  (5) The maximum usable doses for cyclamic acid and its Na and Ca salts are expressed in terms of the free acid
  (6) The maximum usable doses for saccharin and its Na, K and Ca salts are expressed in terms of the free imide

Tryptophan

Regulation: Tryptophan in Food Regulations 1990 (1990/1728)

Definition
Tryptophan: dextrorotatory tryptophan, laevorotatory tryptophan, or racemic tryptophan, or any salt or peptide prepared from any of these forms.

General points
  1) The addition of tryptophan to food intended for sale for human consumption and the sale of food containing tryptophan (where the tryptophan has been added) are prohibited.
  2) Exemptions are provided where there is an appropriate medical certificate indicating that a person requires food with added tryptophan.

4.3

4.3 Contaminants

Aflatoxin

Regulation: Aflatoxins in Nuts, Nut Products, Dried Figs and Dried Fig Products
Regulations 1992 (1992/3236)
Amendment: Food Labelling Regulations 1996 (1996/1499)

Definitions
Aflatoxins: all or any of aflatoxin B_1, aflatoxin B_2, aflatoxin G_1 and aflatoxin G_2.
Total aflatoxins: the sum of the concentrations of aflatoxins.
Nuts: the following nuts (based on EC combined nomenclature subheadings): coconuts (not including desiccated coconut), Brazil nuts (in shell/shelled), cashew nuts, almonds (bitter/other; in shell/shelled), hazelnuts or filberts (*Corylus* spp.) (in shell/shelled), walnuts (in shell/shelled), chestnuts (*Castanea* spp.), pistachios, pecans, areca (or betel) and cola nuts, other nuts and peanuts (not roasted or otherwise cooked (in shell/shelled/broken)/prepared or preserved in immediate packings below 1 kg/greater than 1 kg).
Nut product (for import controls below):   any food at least half of which consists (by weight) of nuts or of any substance (except where exclusively edible oil) derived from nuts.
Nut product (for consumer sale controls below):   any food consisting of any quantity of nuts or of any substance (including edible oil) derived from nuts, and edible oils derived wholly or partly from nuts.
Dried fig product (for import controls below):   any food at least half of which consists (by weight) of dried figs.
Dried fig product (for consumer sale controls below):   any food consisting of any quantity of dried figs.

Import controls
1) Importers of nuts, nut products, dried figs or dried fig products into Great Britain from outside the EEC shall inform the authorities of a consignment of one of these by giving them a certificate detailing what constitutes the consignment. It must be no more than 25 000 kg and must be of one type (i.e. nuts, nut products, dried figs or dried fig products).
2) The import must be through an authorised place of entry and according to detailed procedures allowing for inspection and sampling by the authorities. Provision is made for the examination to be deferred to another place.
3) Sampling procedures specify the number of samples and the size of each sample (depends upon product – see Regulations) and require

4.3

that samples are taken as randomly as possible from throughout the consignment. Preparation of subsamples is detailed. Analysis may be by any method which meets specified performance parameters.

4) The following actions are specified:

a) If total aflatoxins more than $10\,\mu g/kg$: either return to the consignor, or use for purpose other than human consumption or destroy.

b) If total aflatoxins more than $4\,\mu g/kg$ but not more than $10\,\mu g/kg$: as (a) but, in addition, the importer may undertake to process the consignment to ensure that, when a consumer sale takes place, the level is below $4\,\mu g/kg$ but, in this case, merely blending or mixing with a consignment with a lower level of aflatoxins is not a permitted process.

Consumer sales

1) No person may sell any nut, nut product, dried fig or dried fig product which, when analysed using a method meeting specified performance parameters, has a level of total aflatoxins more than $4\,\mu g/kg$.

2) If a sale is found to exceed the limit, it is a defence to show that the nuts or figs concerned came from a lot or batch which had been certified (following the specified sampling and analysis methods) as containing no more than $4\,\mu g/kg$ nuts or dried figs. The laboratory used must be participating in a proficiency testing scheme meeting internationally recognised protocols and must have achieved certain specified results in the scheme.

General point

The Regulations only apply to any nuts, nut products, dried figs or dried fig products which are intended for sale for human consumption.

Arsenic

Regulation: Arsenic in Food Regulations 1959 (1959/831)
Amendments: Arsenic in Food (Amendment) Regulations 1960 (1960/2261)
Arsenic in Food (Amendment) Regulations 1973 (1973/1052)
Colouring Matter in Foods Regulations 1973 (1973/1340)
Emulsifiers and Stabilisers in Food Regulations 1975 (1975/1486)
Flavourings in Food Regulations 1992 (1992/1971)
Bread and Flour Regulations 1995 (1995/3202)

The permitted level for arsenic in foods is:

a) unspecified foods, max. 1 ppm;

b) specified foods, see table under 'Trace metals'.

4.3

General points

1) Where a specified food is used as a component of an unspecified food, in a quantity not less than 25%, the unspecified food may exceed 1 ppm having regard to the quantity of the specified food that is used and the maximum arsenic that is permitted in the specified food.

2) For fish, edible seaweed or products containing these, where arsenic exceeds 1 ppm naturally, arsenic may exceed 1 ppm by virtue of its fish or edible seaweed content.

3) The Regulations do not apply to hops or hop concentrates intended for use for commercial brewing, or to foods where limits are specified in other regulations.

Chloroform

Regulation: Chloroform in Food Regulations 1980 (1980/36)

Added chloroform
No-one may sell or import any food which has in it or on it any added chloroform.

Erucic acid

Regulation: Erucic Acid in Food Regulations 1977 (1977/691)
Amendment: Erucic Acid in Food (Amendment) Regulations 1982 (1982/264)

Note: Erucic acid is not strictly a contaminant being a normal constituent of some foods.

Restrictions
Oil or fat (or mixture):   erucic acid content should not exceed 5% of the fatty acid content.
Any food with added oil or fat: erucic acid content of all the fat and oil in the food should not exceed 5% of the fatty acid content.

Exemptions
The following are exempt from the above standards:
a) any food, except food for infants or young children, containing not more than 5% oil or fat;
b) any oil, fat or food sold to a manufacturer (for manufacturing) or caterer (for catering purposes).

Hormonal substance and other residues in meat

Regulation: Animals, Meat and Meat Products (Examination for Residues and Maximum Residue Limits) Regulations 1991 (1991/2843)
Amendments: Animals, Meat and Meat Products (Examination for Residues and

Maximum Residue Limits) Regulations 1993 (1993/990)
Animals, Meat and Meat Products (Examination for Residues and Maximum
Residue Limits) Regulations 1994 (1994/2465)

4.3

Note: These Regulations are supported by related sections of certain other
Regulations. The area is complex and difficult to summarise.

Definitions

Animal: any of the following food sources, namely animals of the bovine
species (including buffalo of species *Bubalus bubalus* and *Bison bison*),
swine, goats, solipeds, camelids, rabbits, deer and birds reared for
human consumption.

Meat: the flesh or other part of an animal suitable for human
consumption.

Transmissible substance: any substance having a pharmacological action
or any conversion product thereof or any other substance which if
transmitted to meat would be likely to be dangerous to human health.

Authorised substance: a transmissible substance the presence of which in
any animal, meat or meat product is permitted by or in implementation
of Community law.

Hormonal substance: any substance within either of the following
categories – (a) stilbenes and thyrostatic substances; (b) substances with
oestrogenic, androgenic or gestagenic action.

Maximum residue limit: in relation to a concentration of an authorised
substance in the tissues or body fluids of an animal or in any meat or
meat product, means (a) for the substances listed in the following table,
the limits specified in the specified part of the animal; (b) for hormonal
substances authorised by the Animals and Fresh Meat (Hormonal
Substances) Regulations 1988, the maximum natural physiological level
for that substance.

Annex IV substance: a substance specified in Annex IV of Council
Regulation 2377/90 as amended; presently the following substances –
nitrofurans, ronidazole, dapsone, chloramphenicol, dimetridazole.

Requirements

1) It is prohibited to sell any meat (whether or not mixed with other
   food) or any meat product in which there is:
   a) any prohibited substance – i.e. a hormonal substance (defined
      above) unless administered by a veterinary surgeon or
      practitioner for certain specified purposes;
   b) an unlicensed substance – i.e. a substance for which there is no
      veterinary medicinal product licence;
   c) beta-agonist (unless it can be demonstrated that it was authorised
      by a veterinary surgeon);
   d) any Annex IV substance (defined above); or

4.3

e) an authorised substance (defined above) at a concentration exceeding the maximum residue limit (defined above and listed below).

2) The Regulations contain details of inspection, analysis (including indicator residues) and other related requirements.

Maximum residue limits

Pharmacologically active substances	Animal species (1)	Maximum residue limit (MRL) (μg/kg, unless otherwise specified)	Target tissue
A Substances subject to MRLs (not provisional)			
Sulphonamides (all substances belonging to the group)	Bovine, ovine, caprine	100	Milk
Benzylepenicillin	All	50 4	Muscle, liver, kidney, fat Milk
Ampicillin	All	50 4	Muscle, liver, kidney, fat Milk
Amoxicillin	All	50 4	Muscle, liver, kidney, fat Milk
Oxacillin	All	300 30	Muscle, liver, kidney, fat Milk
Cloxacillin	All	300 30	Muscle, liver, kidney, fat Milk
Dicloxacillin	All	300 30	Muscle, liver, kidney, fat Milk
Penethamate	All	50 4	Muscle, liver, kidney, fat Milk
Cefquinome	Bovine	200 100 50 50	Kidney Liver Muscle Fat
Enrofloxancin	Bovine, porcine, poultry	30	Muscle, liver, kidney
Sarafloxacin	Chicken	100 10	Liver Skin + fat
Tilmicosin	Bovine	1000 50	Liver, kidney Muscle, fat
	Ovine Porcine Ovine	1000 50 50 μg/l	Liver, kidney Muscle, fat Milk
Spiramycin	Bovine	300 200 200	Liver, kidney, fat Muscle Milk

(continued)

Pharmacologically active substances	Animal species (1)	Maximum residue limit (MRL) (µg/kg, unless otherwise specified)	Target tissue
	Chicken	400	Liver
		300	Fat + skin
		200	Muscle
Florfenicol	Bovine	200	Muscle
		300	Kidney
		3000	Liver
Tetracycline	All	600	Kidney
		300	Liver
		100	Muscle
		100	Milk
		200	Eggs
Oxytetracycline	All	600	Kidney
		300	Liver
		100	Muscle
		100	Milk
		200	Eggs
Chlortetracycline	All	600	Kidney
		300	Liver
		100	Muscle
		100	Milk
		200	Eggs
Ivermectin	Bovine	100	Liver
		40	Fat
	Ovine, porcine, equidae	15	Liver
		20	Fat
Abamectin	Bovine	20	Liver
		10	Fat
Doramectin	Bovine	15	Liver
		25	Fat
Closantel	Bovine	1000	Muscle, liver
		3000	Kidney, fat
	Ovine	1500	Muscle, liver
		5000	Kidney
		2000	Fat
Levamisole	Bovine, ovine, porcine, poultry	10	Muscle, kidney, fat
		100	Liver
Diazinon	Bovine, ovine, caprine, porcine	700	Fat
		20	Kidney, liver, muscle
	Bovine, ovine, caprine	20	Milk
Carazolol	Porcine	25	Liver, kidney
		5	Muscle, fat + skin

4.3 *(continued)*

Pharmacologically active substances	Animal species (1)	Maximum residue limit (MRL) (µg/kg, unless otherwise specified)	Target tissue
B Substances subject to provisional MRLs (to date indicated)			
Trimethoprim	All	50	Muscle, liver, kidney, fat, milk
Spiramycin (1 July 1997)	Porcine	600	Liver
		300	Kidney, muscle
		200	Fat
Tylosin	Bovine, porcine, poultry	100	Muscle, liver, kidney
(1 July 1997)	Bovine	50	Milk
Erythromycin (1 July 2000)	Bovine, ovine, porcine, poultry	400	Liver, kidney, muscle, fat
	Bovine, ovine	40	Milk
	Poultry	200	Eggs
Josamycin	Chicken	400	Kidney
		200	Liver, muscle, fat
		200	Eggs
Ceftifur (1 July 1997)	Bovine	2000	Kidney, liver
		200	Muscle
		600	Fat
		100	Milk
	Porcine	4000	Kidney
		3000	Liver
		500	Muscle
		600	Fat
Spectinomycin (1 July 1996)	Bovine, porcine, poultry	5000	Kidney
		2000	Liver
		300	Muscle
		500	Fat
	Bovine	200	Milk
Danofloxacin (1 July 1997)	Bovine	900	Liver
		500	Kidney
		300	Muscle
		200	Fat
	Chicken	1200	Liver, kidney
		600	Fat + skin
		300	Muscle
Decoquinate	Bovine, ovine	500	Muscle, liver, kidney, fat
Colistin	Bovine, ovine, porcine, chicken	200	Kidney
	rabbits	150	Liver, muscle, fat
	Bovine, ovine	50	Milk
	Chicken	300	Eggs

(continued)

4.3

Pharmacologically active substances	Animal species (1)	Maximum residue limit (MRL) (µg/kg, unless otherwise specified)	Target tissue
Febantel, fenbendazole, oxfendazole, triclabendazole (1 July 1997)	All	1000 (2)	Liver
		10 (2)	Muscle, kidney, fat
		10 (2)	Milk
	Bovine, ovine	150 (3)	Muscle, liver, kidney
		50 (3)	Fat
Netobimin (1 July 1997)	Bovine, ovine, caprine	1000	Liver
		500	Kidney
		100	Muscle, fat
		100	Milk
Moxidectin (1 July 1997)	Bovine	200	Fat
	Ovine	20	Kidney, liver
Dexamethasone	Bovine, porcine, equidae	2.5	Liver
		0.5	Muscle, kidney
	Bovine	0.3	Milk

Note: (1) All animals for food processing

Lead

Regulation: Lead in Food Regulations 1979 (1979/1254)
Amendments: Lead in Food (Amendment) Regulations 1985 (1985/912)
  Flavourings in Food Regulations 1992 (1992/1971)
  Colours in Food Regulations 1995 (1995/3124)
  Food (Miscellaneous Revocations and Amendment) Regulations 1995 (1995/3267)

The permitted level for lead in foods is:
  a) Unspecified foods, 1.0 mg/kg;
  b) Specified foods, see table under 'Trace metals'.

General points
  1) Unspecified compound foods may contain more than 1 mg/kg if
    a) not less than 10% of the food is an ingredient for which a higher level is permitted, and
    b) if the components were separate, the specified standards would be met.
  2) Additive premixes (marked as not being for retail sale) may be sold above the 1 mg/kg level if the components, if sold separately, would meet the specified standards.
  3) The Regulations do not apply to foods where limits are specified in other regulations.

4.3 Pesticides

Regulation: Pesticides (Maximum Residue Levels in Crops, Food and Feedingstuffs)
Regulations 1994 (1994/1985)
Amendments: Pesticides (Maximum Residue Levels in Crops, Food and
Feedingstuffs) Regulations 1995 (1995/1483)
Pesticides (Maximum Residue Levels in Crops, Food and Feedingstuffs)
(Amendment) Regulations 1996 (1996/1487)

The Regulations specify certain maximum residue levels (MRLs) on certain
foods. Their content is very detailed and it is not appropriate to simplify
the requirements. Readers who need the actual figures are recommended to
obtain copies of the Regulations listed above. Given below are indications
of the pesticides covered and the food categories used in the Regulations.
Not all pesticide/food combinations are covered by specified limits.

Pesticides

Acephate
Aldicarb
Aldrin and dieldrin
2-Aminobutane
Aminotriazole
Amitraz
Atrazine
Azinphos-methyl
Benalaxyl
Benfuracarb
Binapacryl
Bitertanol
Bromophos-ethyl
Bromopropylate
Camphechlor (Toxaphene)
Captafol
Captan
Carbaryl
Carbendazim, benomyl
  and thiophanate-methyl
Carbofuran
Carbon disulphide
Carbon tetrachloride
Carbophenothion
Carbosulphan
Cartap
Chlordane
Chlorfenvinphos
Chlorothalonil
Chlorobenzilate
Chlorpyrifos
Chlorpyrifos-methyl
Cyfluthrin
Cypermethrin
DDT
Daminozide

Deltamethrin
Diazinon
1,2-Dibromoethane
Dichlofluanid
Dichlorvos
Dichlorprop
Dicofol
Diflubenzuron
Dimethipin
Dimethoate
Dinoseb
Dioxathion
Endosulfan
Endrin
Ethephon
Ethion
Etrimfos
Fenarimol
Fenchlorphos
Fenitrothion
Fenvalerate
Fluazifop
Flucythrinate
Flurochloridone
Furathiocarb
Glyphosate
Haloxyfop
Hexachlorobenzene (HCB)
Hexachlorocyclohexane
  (HCH) (as (a) sum of α
  and β, and (b) γ)
Heptachlor
Hydrogen cyanide
Hydrogen phosphate
Imazalil
Inorganic bromide

Ioxynil
Iprodione
λ-Cyhalothrin
Malathion
Maleic hydrazide
Maneb, mancozeb,
  metiram, propineb and
  zineb
Mercury compounds
Metalaxyl
Methacrifos
Methamidophos
Methidathion
Methomyl thiodicarb
Methyl bromide
Mevinphos
Omethoate
Paraquat
Parathion
Parathion-methyl
Permethrin
Phosalone
Phosphamidon
Pirimiphos-methyl
Procymidone
Profenophos
Propiconazole
Pyrethrins
Quintozene
Tecnazene
TEPP
Thiabendazole
Triazophos
Trichlorfon
2,4,5-T
Vinclozolin

Foods

1. Fruit, fresh, dried or uncooked, preserved by freezing not containing added sugars; nuts	Citrus fruit	Grapefruit, lemons, limes, mandarins (including clementines and similar hybrids), oranges, pomelos, others
	Tree nuts (shelled or unshelled)	Almonds, Brazil nuts, cashew nuts, chestnuts, coconuts, hazelnuts, macadamia nuts, pecans, pine nuts, pistachios, walnuts, others
	Pome fruit	Apples, pears, quinces, others
	Stone fruit	Apricots, cherries, peaches (including nectarines and similar hybrids), plums, others
	Berries and small fruit	(a) Table grapes, wine grapes (b) Strawberries (other than wild) (c) Cane fruit: blackberries, dewberries, loganberries, raspberries, others (d) Other small fruit and berries (other than wild): bilberries, cranberries, currants (red, black and white), gooseberries, others (e) Wild berries and wild fruit
	Miscellaneous fruit	Avocados, bananas, dates, figs, kiwi fruit, kumquats, lychees, mangoes, olives, passion fruit, pineapples, pomegranates, others
2. Vegetables, fresh or uncooked, frozen or dry	Root and tuber vegetables	Beetroot, carrots, celeriac, horseradish, Jerusalem artichokes, parsnips, parsley root, radishes, salsify, sweet potatoes, swedes, turnips, yams, others
	Bulb vegetables	Garlic, onions, shallots, spring onions, others
	Fruiting vegetables	(a) Solanacea: tomatoes, peppers, aubergines, others (b) Curcurbits – edible peel: cucumbers, gerkins, courgettes, others (c) Curcurbits – inedible peel: melons, squashes, watermelons, others (d) Sweet corn

4.3

	Brassica vegetables	(a) Flowering brassicas: broccoli, cauliflower, other (b) Head brassicas: Brussels sprouts, headcabbage, others (c) Leafy brassicas: Chinese cabbage, kale, others (d) Kohlrabi
	Leaf vegetables and fresh herbs	(a) Lettuce and similar: cress, lamb's lettuce, lettuce, scarole, others (b) Spinich and similar: Beet leaves (chard) (c) Watercress (d) Witloof (e) Herbs: chervil, chives, parsley, celery leaves, others
	Legume vegetables (fresh)	Beans (with pods), beans (without pods), peas (with pods), peas (without pods), others
	Stem vegetables	Asparagus, cardoons, celery, fennel, globe artichokes, leeks, rhubarb, others
	Fungi	(a) Cultivated mushrooms (b) Wild mushrooms
3. Pulses		Beans, lentils, peas, others
4. Oilseeds		Linseed, peanuts, poppy seeds, sesame seed, sunflower seed (with shell), rape seed, soya bean, mustard seed, cotton seed, others
5. Potatoes		Early potatoes, ware potatoes
6. Tea		(Black tea, processed from the leaves of *Camellia sinensis*)
7. Hops (dried)		Including hop pellets and unconcentrated powder
8. Cereals		Wheat, rye, barley, oats, triticale, maize, rice (including paddy or rough rice, husked rice and semi-milled or wholly milled rice), other cereals
9. Products of animal origin		Meat, fat and preparations of meat Milk and dairy produce Eggs
10. Spices		Cumin seed, juniper berries, nutmeg, pepper (black and white), vanilla pods, others

General points
1) The Regulations specify maximum levels of certain pesticide residues which may be left in certain products after the application to that

146

4.3

product (crop, food or feedingstuff) or to land on which it is grown.
2) No person may leave or cause to be left in any of the foods specified in the Regulations a level of residue of the pesticide exceeding that specified (where one is specified in the Regulations).
3) The substances which are regarded as comprising the specified residues are listed in the Regulations.
4) Levels of residues should be determined (as far as practicable) using procedures recommended in publications of the Codex Alimentarius.

Radioactive contamination

Regulations: Commission Regulations (Euratom) 3954/87, 944/89
Amendments: Commission Regulation (Euratom) 2218/89

These controls were introduced as a response to the Chernobyl accident but the levels have been retained so as to allow rapid implementation in the event of any future incident.

Maximum permitted levels for foodstuffs and feedingstuffs
The following values (in Bq/kg) have been established by the Comission for foodstuffs (1):

	Baby foods (2)	Dairy produce (3)	Other foodstuffs except minot foodstuffs (4)	Liquid foodstuffs (5)
Isotopes of strontium, notably ^{90}Sr	75	125	750	125
Isotopes of iodine, notably ^{131}I	150	500	2000	500
Alpha-emitting isotopes of plutonium and transplutonium elements, notably ^{239}Pu, ^{241}Am	1	20	80	20
All other nuclides of half-life greater than 10 days, notably ^{134}Cs, ^{137}Cs (6)	400	1000	1250	1000

Notes (as taken from the Commission Regulation 3954/87, as amended):
(1) The level applicable to concentrated or dried products is calculated on the basis of the reconstituted product as ready for consumption. Member States may make recommendations concerning the diluting conditions in order to ensure that the maximum permitted levels laid down in this regulation are observed
(2) Baby foods are defined as those foodstuffs intended for the feeding of infants during the first 4 to 6 months of life, which meet, in themselves, the nutritional requirements of this category of person and are put up for retail sale in packages which are clearly identified and labelled 'food preparation for infants'
(3) Dairy produce is defined as those products falling within the following cn codes including, where appropriate, any adjustments which might be made to them later: 0401, 0402 (except 0402 29 11)

147

4.3

(4) Minor foodstuffs and the corresponding levels to be applied to them have been defined (see below)

(5) Liquid foodstuffs as defined in the heading 2009 and in Chapter 22 of the combined nomenclature. Values are calculated taking into account consumption of tap-water and the same values should be applied to drinking water supplies at the discretion of competent authorities in Member States

(6) Carbon-14, tritium and potassium-40 are not included in this group

Minor foodstuffs

For the following minor foodstuffs, the maximum permitted levels to be applied are 10 times those applicable to 'other foodstuffs except minor foodstuffs' given above:

Garlic (fresh or chilled).

Truffles (fresh or chilled).

Capers (fresh or chilled).

Capers (provisionally preserved, but unsuitable in that state for immediate consumption).

Truffles (dried, whole, cut, sliced, broken or in powder, but not further prepared).

Manioc, arrowroot, salep, Jerusalem artichokes, sweet potatoes and similar roots and tubers with high starch or inulin content, fresh or dried, whether or not sliced or in the form of pellets; sago pith.

Peel of citrus fruit or melons (including watermelons), fresh, frozen, dried or provisionally preserved in brine, in sulphur water or in other preservative solutions.

Maté.

Pepper of the genus *Piper*; dried or crushed ground fruits of the genus *Capsicum* or of the genus *Pimenta.*

Vanilla.

Cinnamon and cinnamon-tree flowers.

Cloves (whole fruit, cloves and stems).

Nutmeg, mace and cardamons.

Seeds of anise, badian, fennel, coriander, cumin or caraway; juniper berries.

Ginger, saffron, turmeric (curcuma), thyme, bay leaves, curry and other spices.

Flour and meal of sago, roots or tubers.

Manioc (cassava) starch.

Hop cones, fresh or dried, whether or not ground, powdered or in the form of pellets; lupulin.

Plants and parts of plants (including seeds and fruits), of a kind used primarily in perfumery, in pharmacy or for insecticidal, fungicidal or similar purposes, fresh or dried, whether or not cut, crushed or powdered.

Lac; natural gums, resins, gum-resins and balsams.

Vegetable saps and extracts; pectic substances, pectinates and pectates; agar-agar and other mucilages and thickeners, whether or not modified, derived from vegetable products.

Fats and oils and their fractions, of fish or marine mammals, whether or not refined, but not chemically modified.

Caviar and caviar substitutes.

Cocoa beans, whole or broken, raw or roasted.

Cocoa shells, husks, skins and other cocoa waste.

Cocoa paste, whether or not defatted.

Truffles (prepared or preserved otherwise than by vinegar or acetic acid).

Fruit, nuts, fruit-peel and other parts of plants, preserved by sugar (drained, glacé or crystallized).

Yeasts (active or inactive); other single-cell micro-organisms, dead (but not including vaccines of heading No. 3002); prepared baking powders.

Provitamins and vitamins, natural or reproduced by synthesis (including natural concentrates), derivatives thereof used primarily as vitamins, and intermixtures of the foregoing, whether or not in any solvent.

Essential oils (terpeneless or not), including concretes and absolutes; resinoids; concentrates of essential oils in fats, in fixed oils, in waxes or the like, obtained by enfleurage or maceration; terpenic byproducts of the deterpenation of essential oils; aqueous distillates and aqueous solutions of essential oils.

4.3

Tin

Regulation: Tin in Food Regulations 1992 (1992/496)

No-one may sell or import any food containing a level of tin exceeding 200 mg/kg.

Trace metals

The standards for the trace metals arsenic and lead in specified foods are as given below:

Food / food ingredient	Statutory limit (ppm, mg/kg)	
	Lead	Arsenic
Angelica, glacé	2.0	
Apples	1.0	
Beer	0.2	
Beer: black beer or black beer and rum		0.5
Beverages: alcoholic, not otherwise specified		0.2

4.3 *(continued)*

Food / food ingredient	Statutory limit (ppm, mg/kg)	
	Lead	*Arsenic*
Beverages: non-alcoholic ready to drink, not otherwise specified	0.2	0.1
Brandy	0.2	
Chemicals: not otherwise specified	10	2.0
Chemicals: for which arsenic limits are specified in the *British Pharmacopoeia* or the *British Pharmaceutical Codex*		(1)
Chemicals: for which lead limits are specified in the *European Pharmacopoeia*, *British Pharmacopoeia* or the *British Pharmaceutical Codex*	(1)	
Chicory: dried and roasted		4.0
Cider	0.2	
Cocoa powder	2.0	
Corned beef	2.0	
Curry powder	10	
Dandelion coffee including soluble dandelion coffee compound	5.0	
Fats: edible	0.5	
Finings and clearing agents		5.0
Fish: not otherwise specified	2.0	
Fish: dried	5.0	
Flavourings	10.0	3.0
Frozen confections (and any other similar commodity)		0.5
Fruit: dried or dehydrated	2.0	
Fruit juices: concentrated (excluding lime or lemon)	1.0	
Fruit juices: ready to drink (excluding lime or lemon)	0.5	
Fruit juices: undiluted		0.5
Fruit juices: concentrates for consumption after dilution (max. 10 parts to 1) for use by manufacturers of soft drinks		0.5
Game; game pâté	10	
Gelatine: edible		2.0
Geneva	0.2	
Gin	0.2	
Herbs: dried	10	5.0
Hops: concentrates, excluding those for commercial brewing		5.0
Hops: dried, excluding those for commercial brewing		2.0
Hops	10	
Ice-cream (including frozen confections and similar commodities)		0.5
Infant food (excluding rusks and dried, dehydrated or concentrated infant food)	0.2	
Iron: iron powder, used in the preparation of flour		10
Lemon/lime juice: concentrated (for manufacturing)	2.0	
Liquorice: dried extract		2.0
Liver	2.0	
Milk: condensed	0.5	
Mustard	10	5.0
Oils: edible	0.5	
Onions: dehydrated		2.0
Pears	1.0	
Perry	0.2	
Pickles and sauces	2.0	

(*continued*)

4.3

Food / food ingredient	Statutory limit (ppm, mg/kg)	
	Lead	*Arsenic*
Proteins: hydrolysed	2.0	
Rum	0.2	
Rusks	0.5	
Shellfish	10	
Soft drink concentrates: for use in the manufacture of soft drinks		0.5
Soft drinks: concentrated		0.5
Soft drinks: ready to drink	0.2	
Spices: including ground spices	10	5.0
Sugars (defined as any soluble carbohydrate sweetening matter):		
a) sulphated ash content max. 0.25%	0.5	
b) sulphated ash content 0.25–1.0%	1.0	
c) sulphated ash content over 1.0% (not for further refining)	5.0	
d) sulphated ash content over 1.0% (for further refining)	10	
Tea	5.0	
Tomato juice, tomato juice cocktails: concentrated	1.0	
Tomato juice, tomato juice cocktails: ready to drink	0.5	
Tomato paste, purée or powder		
a) total solids 15–25%	2.0	
b) total solids not less than 25%	3.0	
Vegetable juices: concentrated	1.0	
Vegetable juices: ready to drink	0.5	
Vegetables: dehydrated or dried but excluding onion	2.0	
Vodka	0.2	
Water ices or similar frozen confection	0.2	
Whisky	0.2	
Yeast: brewers' yeast for the manufacture of yeast products		5.0 (3)
Yeast and yeast products: but excluding yeast extract or, for arsenic, brewers' yeast for the manufacture of yeast products	5.0 (3)	2.0 (3)
Yeast extract	2.0	

Notes:
  (1)   As specified in named publications
  (2)   Calculated on the dry fat-free substance
  (3)   Calculated on dry matter

4.4 Processing and packaging

Frozen foods

Regulation: Quick-frozen Foodstuffs Regulations 1990 (1990/2615)
Amendments: Quick-frozen Foodstuffs (Amendment) Regulations 1994 (1994/298)
  Food Labelling Regulations 1996 (1996/1499)

Note: These Regulations only apply to a frozen food if it falls within the
definition of a 'quick-frozen foodstuff'.

4.4

Definition

Quick-frozen foodstuff: a product (a) comprising food which has undergone a freezing process known as 'quick-freezing' whereby the zone of maximum crystallisation is crossed as rapidly as possible, depending on the type of product, and (b) which is labelled for the purpose of sale to indicate that it has undergone that process.

Specified processing conditions

1) A quick-frozen foodstuff must meet the following conditions:
   a) it has been manufactured from raw materials of sound, genuine and merchantable quality and no other materials;
   b) no raw material has been used in its manufacture unless, at the time of its use, it would have been lawful for it to be sold for human consumption;
   c) its preparation and quick-freezing have been carried out with sufficient promptness, and by the use of appropriate technical equipment, to minimise any chemical, biochemical and microbiological changes to the food comprised in it;
   d) only authorised cryogenic media may be used in direct contact with any food contained in it (see (2) below);
   e) the quick-freezing of each food contained in it has resulted in the temperature of that food after thermal stabilisation being −18°C or colder; and
   f) following the quick-freezing and thermal stabilisation of any food to which (e) applies the temperature of that food has been maintained or has, save during the application of any one of the permitted exceptions, been maintained at a level or levels no warmer that −18°C (see (3) below).
2) Authorised cryogenic media are: air, nitrogen and carbon dioxide.
3) The following temperature exceptions are permitted:
   a) brief periods during transport other than local distribution – warmer than −18°C but not warmer than −15°C;
   b) during local distribution (delivery to the point of retail sale or to a catering establishment) – warmer than −18°C (to an extent consistent with good distribution practice) but not warmer than −15°C;
   c) in a retail display cabinet (after 9 January 1997) – warmer than −18°C to an extent consistent with good storage practice but not warmer than −12°C.

Packaging

Prepackaging used by a manufacturer or packer must be suitable to protect the food from microbial and other forms of external contamination and against dehydration, and the food must remain in the prepackaging until sale.

4.4

Labelling
1) The description 'quick-frozen' may only be used for quick-frozen foodstuffs.
2) Quick-frozen foods for sale to the ultimate consumer (or catering establishment) shall be marked or labelled with:
   a) date of minimum durability,
   b) maximum advised storage period,
   c) advised storage temperature and/or storage equipment,
   d) reference to the batch to which it belongs, and
   e) a message of the type 'do not refreeze after defrosting'.
3) Quick-frozen foods not included in (2) shall be marked or labelled with:
   a) reference to the batch to which it belongs, and
   b) name and address of manufacturer or packer or seller within the EEC.

Recording instruments
1) Each manufacturer, storer, transporter, local distributor and retailer shall ensure that any storage or transport (excluding rail transport) used is fitted with specified instruments for measuring or monitoring of the air temperature (see Regulations for specified requirements).
2) Each manufacturer, storer or transporter (excluding local distribution) shall at frequent and regular intervals record the air temperatures, shall date such records and keep them for 1 year (or longer if justified by nature of foodstuff) and shall make records available if requested.

General points
1) Each manufacturer, storer, transporter, local distributor and retailer shall ensure that equipment used for quick-frozen foodstuffs does not lead to a contravention of the Regulations.
2) If an inspector has doubts as to the temperatures of the food, he shall further inspect it according to procedures specified in Directive 92/2.

Extraction solvents

Regulation: Extraction Solvents in Food Regulations 1993 (1993/1658)
Amendment: Extraction Solvents in Food (Amendment) Regulations 1995 (1995/1440)

Definitions
Extraction solvent: any solvent which is used or intended to be used in an extraction procedure and includes in any particular case, further to its use in such a procedure, any substance other than such a solvent but deriving exclusively from such a solvent.

153

4.4

Solvent: any substance which is capable of dissolving food, or any ingredient or other component part of food, including any contaminant which is in or on a food.

Residue: includes any residual substance other than an extraction solvent but deriving exclusively from an extraction solvent.

Permitted extraction solvents

1) The following may be generally used as extraction solvents: water (whether or not other substances regulating acidity or alkalinity have been added to it), any food possessing solvent properties.

2) The following extraction solvents may be generally used so long as their use has resulted in the presence in or on the food only of residues in technically unavoidable quantities that present no danger to human health: propane, butane, butyl acetate, ethyl acetate, ethanol, carbon dioxide, acetone (not to be used for olive–pomace oil), nitrous oxide.

3) The following extraction solvents may be generally used so long as its (their) use results in the presence of a maximum residue (singly or combined) of 10 mg/kg: methanol, propan-2-ol.

4) The following extraction solvents may only be used for the purposes listed in the following table and must result in a residue no greater than the maximum specified: hexane, methyl acetate, ethyl methyl ketone, dichloromethane (but note that hexane and ethyl methyl ketone are not to be used in combination).

Specified foods	Permitted extraction solvents and purpose for addition	Maximum residue of solvent in the food (or as specified)
Fats	a) Hexane, for the production or fractionation of the fats; or b) Ethyl methyl ketone for the fractionation of the fats	5 mg/kg
Oils	a) Hexane, for the production or fractionation of the oils; or b) Ethyl methyl ketone for the fractionation of the oils	5 mg/kg
Cocoa butter	Hexane, for the prodution of cocoa butter	5 mg/kg
Protein products	Hexane, for the production of protein products	10 mg/kg in any food in which the protein products are an ingredient
Defatted flours	Hexane, for the preparation of defatted flours	10 mg/kg in any food in which the defatted flours are an ingredient
Defatted cereal germs	Hexane, for the preparation of defatted cereal germs	5 mg/kg

4.4

(*continued*)

Specified foods	Permitted extraction solvents and purpose for addition	Maximum residue of solvent in the food (or as specified)
Defatted soya product	Hexane, for the preparation of defatted soya products	30 mg/kg in the defatted soya product as sold to the final consumer
Coffee	Methyl acetate, ethyl methyl ketone or dichloromethane (alone or in combination), for decaffeination and/or removal of irritants and bitterings	a) Methyl acetate or ethyl methyl ketone, 20 mg/kg (1) b) Dichloromethane, 2 mg/kg
Tea	Methyl acetate, ethyl methyl ketone or dichloromethane (alone or in combination), for decaffeination and/or removal of irritants and bitterings	a) Methyl acetate or ethyl methyl ketone, 20 mg/kg (1) b) Dichloromethane, 5 mg/kg
Sugar from molasses	Methyl acetate, for production of the sugar from molasses	1 mg/kg

Note: (1)   If used in combination, the combined total residue shall not exceed 20 mg/kg

5) The following extraction solvents may be used for the preparation of flavourings prepared from natural flavouring materials provided that, as a result of their use, any food containing or consisting of the flavourings has no more residue than indicated: hexane (1 mg/kg), methyl acetate (1 mg/kg), ethyl methyl ketone (1 mg/kg), dichloromethane (0.02 mg/kg), diethyl ether (2 mg/kg), butan-1-ol (1 mg/kg), butan-2-ol (1 mg/kg), methyl propan-1-ol (1 mg/kg), propan-1-ol (1 mg/kg), cyclohexane (1 mg/kg).

General
1) The Regulations do not apply to extraction solvents used in the production of any food additive, vitamin, or any other nutritional additive except where specified above.
2) The Regulations also contain requirements for the labelling of permitted extraction solvents.

Irradiation

Regulation: Food (Control of Irradiation) Regulations 1990 (1990/2490)

Definition
Ionising radiation: any γ-rays, X-rays or corpuscular radiations which are capable of producing ions either directly or indirectly other than:
(a) those rays or corpuscular radiations: which are emitted by measuring or inspection devices; which are emitted at an energy level no higher

4.4

than 10 MeV in the case of X-rays and 5 MeV otherwise; and the dose of energy imparted by which does not exceed 0.5 Gy; and
(b) those rays or corpuscular radiations applied in respect of food prepared under medical supervision for patients requiring sterile diets.

Only the following foods can be treated with ionising radiation to the maximum dose indicated:

Food description	Maximum overall average dose (kGy) (1)
Fruit (including fungi, tomatoes and rhubarb)	2
Vegetables (excluding fruit, cereals, bulbs and tubers and spices and condiments but including pulses)	1
Cereals (2)	1
Bulbs and tubers (meaning potatoes, yams, onions, shallots and garlic)	0.2
Spices and condiments (meaning dried substances normally used for seasoning)	10
Fish and shellfish (including eels, crustaceans and molluscs)	3
Poultry (meaning domestic fowls, geese, ducks, guinea fowls, pigeons, quails and turkeys)	7

Notes:
(1) An additional requirement states that the maximum dose must be the lower of either (a) 3 times the minimum dose, or (b) 1.5 times the level stated in this table
(2) Cereals are defined by reference to an EC Council Regulation

General points
1) No person shall treat food for sale by ionising radiation unless he holds a current irradiation licence, he is permitted by the licence to irradiate the food and the irradiation complies with the conditions of the licence. Full details concerning the issuing of licences are given in the Regulations.
2) No person shall import irradiated food, whether or not an ingredient of another food, unless it is permitted (see table above), is of recognised appropriate origin and is accompanied by appropriate documentation (see Regulations for full details).
3) No person shall store or transport irradiated food, whether or not an ingredient of another food, unless accompanied by specified documentation.
4) No person shall sell irradiated food, or food containing irradiated food, unless either (1) or (2) were observed and (3) was observed.
5) Food for export to the EC may be treated, stored or transported only if it complies with the legislation of the member state concerned.

Materials and articles in contact with food

A General

Regulation: Materials and Articles in Contact with Food Regulations 1987 (1987/1523)

Amendment: Materials and Articles in Contact with Food (Amendment) Regulations 1994 (1994/979)

No person shall sell, import or use in connection with food any material which does not satisfy the following standards.

General points

1) Materials and articles for food use shall be manufactured in such a way that under normal or foreseeable conditions of use they do not transfer their constituents to food in quantities which could:
   a) endanger human health, or
   b) bring about a deterioration in the organoleptic characteristics of such food or an unacceptable change in its nature, substance or quality.
2) The following particulars are required when any food contact material and article is sold in its finished state:
   a) the description 'for food use', or a specific indication of the particular use for the material or article, or the designated symbol for food contact materials (see below) (for retail sale this provision does not apply if the material or article is by its nature clearly intended to come into contact with food),
   b) any special conditions to be observed in use,
   c) the name and address or trade mark of the producer or a seller within the EEC.

4.4

Vinyl chloride monomer
Materials manufactured with vinyl chloride polymers or copolymers shall not contain vinyl chloride monomer in a quantity exceeding 1 mg/kg of the material. In addition, they should be manufactured so that they do not transfer vinyl chloride to food in which they are in contact at a level exceeding 0.01 mg/kg of food.

Regenerated cellulose film
1) Regenerated cellulose film is defined as: a thin-sheet material obtained from refined cellulose derived from unrecycled wood or cotton, with or without the addition of suitable substances, either in the mass or on one or both surfaces.
2) Controls are placed on film intended to come into contact with food (or is in contact with food and intended for that purpose) and which (a) constitutes a finished product in itself, or (b) is a part of a finished product containing other materials. However, excluded is film which (a) has a coating exceeding 50 mg/dm^2 on the side intended to come into contact with food, and (b) synthetic casings of regenerated cellulose.
3) The following specific provisions relate to regenerated cellulose film:
   a) It must be manufactured from a list of specified substances (see Regulations).
   b) It must be manufactured so that it does not transfer any detectable quantity of adhesive or colourant to food.
   c) If coated, it must not transfer bis(2-hydroxyethyl) ether or ethanediol (or both these substances) in excess of 30 mg/kg food.
   d) In the course of business (involving storage, preparation, packaging, selling or serving food):
      – regenerated cellulose film containing bis(2-hydroxyethyl) ether or ethanediol (or both) shall not be used for food containing water physically free at the surface;
      – no printed surface of regenerated cellulose film shall come into contact with food.

B Plastic
Regulation: Plastic Materials and Articles in Contact with Food Regulations 1992 (1992/3145)
Amendments: Plastic Materials and Articles in Contact with Food (Amendment) Regulations 1995 (1995/360)
Plastic Materials and Articles in Contact with Food (Amendment) Regulations 1996 (1996/694)

Note: Plastic materials are subject to the general controls described above in addition to these more specific controls.

Transfer of constituents

Maximum overall migration limits are specified as follows:

4.4

a) In the case of containers (capacity 500 ml to 10 l), articles for filling where the surface cannot be estimated, and caps, gaskets, stoppers or similar sealing devices: 60 mg/kg of food

b) In any other case: 10 mg/dm^2 of plastic surface area.

Monomers

The Regulations specify the permitted monomers and place restrictions on the use of some of them (i.e. quantity in the plastic material or maximum specific migration limits). Certain monomers are only permitted to be used until 31 December 1996 (See Regulations). The following are exempt from this part of the Regulations: surface coatings obtained from resinous or polymerized products in liquid, powder or dispersion form, including varnishes, lacquers and paints; silicones; epoxy resins; products obtained by means of bacterial fermentation; adhesives and adhesion promoters; printing inks.

Simulants

For the purposes of assessing whether plastic materials transfer constituents to food, the Regulations specify, in some cases, simulants appropriate for various foods (see Regulations). The simulants specified are:

Simulant A: distilled water or water of equivalent quality.
Simulant B: 3% acetic acid (w/v) in aqueous solution.
Simulant C: 15% ethanol (v/v) in aqueous solution.
Simulant D: rectified olive oil or, if technically necessary, a mixture of synthetic triglycerides or sunflower oil (as defined in the Regulations).

If these are not appropriate, other simulants may be used.

Where no simulant is specified but the food is listed, the food should be used; where a food is not listed, the simulant which corresponds most closely in extractive capacity should be used.

4.5 Labelling

General labelling requirements

Regulation: Food Labelling Regulations 1996 (1996/1499)

Notes: The MAFF have issued some Notes For Guidance to help in the interpretation of these Regulations. However they have no status in law and the details that follow have made no reference to them. It is

4.5 recommended though that people needing to operate these Regulations should obtain a copy of the latest issue of the notes.

A Food to be delivered to the ultimate consumer or caterer

1) Exemptions (but in the case of (a) and (b), the foods must carry statements relating to packaging gases and sweeteners – see Additional statements below):

   a) The following foods are not subject to the controls of this section: specified sugar products, cocoa and chocolate products, honey, condensed and dried milk for delivery to a catering establishment (unless prepared and labelled for infant consumption), coffee and coffee products (including designated chicory products) for delivery to a catering establishment.

   b) The following foods are not subject to the controls of this section insofar as their labelling is regulated by other Regulations: hen eggs, spreadable fats, wines or grape musts, sparkling wines and aerated sparkling wines, liqueur wines, semi-sparkling wines and aerated semi-sparkling wines, spirit drinks, fresh fruit and vegetables, preserved sardines, preserved tuna and bonito, additives.

   c) The following foods are not subject to the Regulations: drinks bottled before 1 January 1983 having an alcoholic strength greater than 1.2% (vol.) and meeting the labelling requirements in force at the time of bottling; any food prepared on domestic premises for sale for the benefit of the person preparing it by a registered society; any food not prepared in the course of a business by the person preparing it.

2) All food to be marked with:

   a) the name of the food;

   b) a list of ingredients;

   c) the appropriate durability indication:

   (i) for food which, from a microbiological point of view, is highly perishable and in consequence likely after a short period of time to constitute an immediate danger to human health, a 'use by' date,

   (ii) for other foods, the minimum durability;

   d) any special storage conditions or conditions of use;

   e) the name and address of the manufacturer or packer, or a seller established with the EC;

   f) particulars of the place of origin if necessary to avoid misleading the purchaser to a material degree;

   g) instructions for use if necessary.

Name of food

If a name prescribed by law exists, it shall be used, and may be qualified

4.5

by other words which make it more precise. (For fish, Schedule 1 of the Regulations should be consulted; for melons and potatoes sold as such, the variety must be included; for vitamins, the name should be that used in the table on page 171 except that folacin may be called 'folic acid' and vitamin K shall be called 'vitamin K').

If no name prescribed by law exists, a customary name may be used.

If no name prescribed by law nor a customary name, a name sufficiently precise to inform a purchaser of the true nature of the food and to enable the food to be distinguished from products with which it could be confused and, if necessary, shall include description of its use.

The name may consist of a name and/or a description (and may contain more than one word).

Trade marks, brand names, or fancy names shall not be substituted.

If the purchaser could be misled without such information, the name should include an indication (a) that a food is powdered or in any other physical condition, or (b) that a food has been dried, freeze-dried, frozen, concetrated or smoked or subjected to any other treatment. In addition, the following indications are required:

a) for meat treated with proteolytic enzymes – 'tenderised',
b) for food which has been irradiated – 'irradiated' or 'treated with ionising radiation'.

List of ingredients

Heading: the list of ingredients should be headed or preceded by 'ingredients' (or a heading which includes the word 'ingredients').

Order of ingredients: ingredients to be listed in weight descending order determined as at the time of their use in the preparation of the food, except for the following:

a) water and volatile products used as ingredients shall be listed in order of their weight in the finished product. The weight of water is calculated by subtracting from the weight of the finished product the total weight of the other ingredients used;
b) if an ingredient is reconstituted from concentrated or dehydrated form during preparation of the food, it may be positioned according to its weight before concentration or dehydration;
c) if a food is to be reconstituted during use by the addition of water, its ingredients may be listed in order after reconstitution provided there is a statement 'ingredients of the reconstituted product' or 'ingredients of the ready to use product' or similar indication;
d) if a product consists of mixed fruit, nuts, vegetables, spices or herbs and no particular one of these ingredients predominates significantly by weight, the ingredients may be listed in any order provided that for foods consisting entirely of such a mixture the heading includes 'in variable proportion' or other words indicating method of listing

4.5

and for other such foods the relevant ingredients are accompanied by such a statement.

Names of ingredients: the name of an ingredient shall be:

a) the name which would be used if the ingredient were sold as a food, including, if appropriate, either 'irradiated' or 'treated with ionising radiation' (other appropriate indications must be given if a consumer could be misled by its omission); or

b) the generic name given in the following table and meeting the specified conditions; or

Generic name	Ingredients	Conditions of use
Cheese	Any type of cheese or mixture of cheese	The labelling of the food of which the cheese is an ingredient must not refer to a specific type of cheese
Cocoa butter	Press, expeller or refined cocoa butter	
Crumbs or rusks, as is appropriate	Any type of crumbed, baked cereal product	
Crystallised fruit	Any crystallised fruit	Max. 10% crystallised fruit in the food of which it is an ingredient
Dextrose	Anhydrous dextrose or dextrose monohydrate	
Fat	Any refined fat	The generic name to include either: a) 'animal' or 'vegetable' as appropriate, or b) a specific animal or vegetable, and, if appropriate, 'hydogenated'.
Fish	Any species of fish	The labelling of the food of which the fish is an ingredient must not refer to a specific species of fish
Flour	Any mixture of flour derived from two or more cereal species	The generic name shall be followed by a list of the cereals in the flour (in weight descending order)
Glucose syrup	Glucose syrup or anhydrous glucose syrup	
Gum base	Any type of gum preparation used in the preparation of chewing gum	
Herb, herbs or mixed herbs	Any one or more herb or parts of a herb	Max. 2% herb or herbs in the food of which it or they are an ingredient
Milk proteins	Any caseins, caseinates or whey proteins (or mixture)	
Oil	Any refined oil, other than olive oil	The generic name to include either:

(*continued*)

Generic name	Ingredients	Conditions of use
		a) 'animal' or 'vegetable' as appropriate, or b) a specific animal or vegetable, and, if appropriate, 'hydogenated'
Spice, spices or mixed spices	Any one or more spices	Max. 2% spices in the food of which it or they are an ingredient
Starch	Any unmodified starch or any starch which has been modified either by physical means or enzymes	
Sugar	Any type of sucrose	
Vegetables	Any mixture of vegetables	Max. 10% vegetables in the food of which it or they are an ingredient
Wine	Any type of wine defined in EEC Regulation 822/87	

c) a flavouring shall be identified by the word 'flavouring' or 'flavourings' or a more specific name or description of the flavouring; it may be supplemented by the word 'natural' (or similar) only where the flavouring component(s) of the ingredient consist(s) exclusively of a flavouring substance obtained by physical (including distillation and solvent extraction), enzymatic or microbiological processes, from material of vegetable or animal origin which material is either raw or has only been subjected to a process normally used in preparing food (including drying, torrefaction and fermentation), and/or a flavouring preparation(s); in addition, if the name refers to a vegetable or animal nature or origin, the word 'natural' (or similar), may only be used when the flavouring component has been isolated solely or almost solely from that vegetable or animal source.

d) an additive shall be listed by either the principal function it serves, as given in the following table, followed by its name and/or serial number (subject to the notes in the table) or, where the function is not given in the table, its name:

Acid (1)	Emulsifying salts	Preservative
Acidity regulator	Firming agent	Propellant gas
Anti-caking agent	Flavour enhancer	Raising agent
Anti-foaming agent	Flour treatment agent	Stabiliser
Antioxidant	Gelling agent	Sweetener
Bulking agent	Glazing agent	Thickener
Colour	Humectant	
Emulsifier	Modified starch (2)	

See overleaf for Notes

163

4.5

Notes:
(1) If used as an acid and the specific name includes the word 'acid', the category name need not been given
(2) Neither the specific name nor the serial number need be indicated

Compound ingredients: the names of the ingredients of a compound ingredient may be given either instead of the compound ingredient, or in addition (and immediately following the name of the compound ingredient); except only the name of the compound ingredient need be given if the compound ingredient:
a) need not bear an ingredients' list if it were being sold,
b) is identified by a generic name, or
c) is less than 25% of the finished product – but in this case, any additives used and needing to be named must be listed immediately following the name of the compound ingredient.

Water: added water to be declared unless:
a) it is used solely for reconstitution of an ingredient which is in concentrated or dehydrated form,
b) it is used as, or as part of, a medium which is not normally consumed,
c) it does not exceed 5% of the finished product,
d) it is permitted under EEC frozen or quick-frozen poultry Regulations.

Ingredients not needing to be named:
a) constituents which are temporarily separated and later reintroduced (in the original proportions),
b) additives which were in an ingredient and which serve no significant technological function in the finished product,
c) any additive used solely as a processing aid,
d) any substance (other than water) used as a solvent or carrier of an additive (and used only at a level which is strictly necessary).

Foods which need not bear a list of ingredients:
a) fresh fruit and vegetables which have not been peeled or cut into pieces;
b) carbonated water (consisting of water and carbon dioxide only, and the name indicates that the water is carbonated);
c) vinegar derived by fermentation (from a single basic product) with no added ingredients;
d) cheese, butter, fermented milk and fermented cream to which only lactic products, enzymes and micro-organism cultures essential to manufacture have been added, or, in the case of cheese (except fresh curd cheese and processed cheese), any salt required for its manufacture;
e) any food consisting of a single component (including flour containing only legally required nutritional additives);
f) any drink with an alcoholic strength by volume over 1.2%.

164

For (c) and (d), if other ingredients are included only those other added ingredients need be listed if the list is headed 'added ingredients' or similar.

4.5

Special emphasis
Where a food is characterised by the presence of a particular ingredient, and special emphasis is given to it on the label, it must include a declaration of the minimum percentage of the ingredient. Similarly, for the low content of a particular ingredient, the maximum percentage should be given. The percentage is to be calculated on the quantity used and to appear by the name of the food or in the ingredients' list. (Neither reference to an ingredient in the name of the food nor to a flavouring on the label shall of itself constitute special emphasis.)

Appropriate durability indication
1) General requirements:
    a) Minimum durability: to be indicated by 'best before' followed by the date up to and including which the food can reasonably be expected to retain its specific properties if properly stored, and details of the storage conditions necessary for the properties to be retained until that date. Date to be in the form day/month/year except day/month only required if date is within 3 months, or 'best before end' and month/year if date is from 3 to 18 months, or 'best before end' and month/year or year if date is more than 18 months. Either the date only or the date and storage conditions may be elsewhere on the packet if reference is made to the position after the 'best before' statement.
    b) 'Use by' date: to be indicated by 'use by' followed by the date (expressed in day/month or day/month/year) up to and including which the food, if properly stored, is recommended for use. Details of any necessary storage conditions must also be given. Either the date only or the date and storage conditions may be elsewhere on the packet if reference is made to the position after the 'use by' statement.
2) Foods exempted from stating an appropriate durability indication:
    a) fresh fruit and vegetables (including potatoes but not including sprouting seeds, legume sprouts and similar products) which have not been peeled or cut into pieces,
    b) wine, liqueur wine, sparkling wine, aromatised wine and any similar drink obtained from fruit other than grapes and certain other drinks made from grapes or grape musts (see Regulations),
    c) any drink with an alcoholic strength by volume of 10% or more,
    d) any soft drink, fruit juice or fruit nectar or alcoholic drink sold in a container of more than 5 l (intended for catering),

4.5

e) flour confectionery and bread normally consumed within 24 hours of preparation,

f) vinegar,

g) cooking and table salt,

h) solid sugar and products consisting almost solely of flavoured or coloured sugars,

i) chewing gums and similar products,

j) edible ices in individual portions.

Omission of certain particulars

a) Food which is not prepacked and fancy confectionery products (includes food which is prepacked for direct sale, flour confectionery packed in crimp case or transparent packaging unmarked or marked only with price and lot mark and no other label is attached or details given, and individually wrapped fancy confectionery products not enclosed in any further packaging and intended for sale as single items):

   i) Subject to (ii), food which is not prepacked need not be marked with any of the above requirements (or the additional labelling requirements (a)–(f) below) except the name of the food. In the case of milk, if required to avoid confusion, the place of origin is to be given, and, if raw milk, the name of manufacturer or packer.

   ii) The following (unless irradiated) need not be marked with any of the above requirements: food which is not exposed for sale, white bread or flour confectionery, carcases and parts of carcases not intended for sale in one piece (but see (b) below).

b) Additives: items in (a) which have no list of ingredients and contain additives normally requiring declaration, and which serve as antioxidant, colour, flavouring, flavour enhancer, preservative or sweetener, must be marked with an indication of any such category of additive in the food. For edible ices and flour confectionary, a notice displayed in a prominent position stating that the products may contain stated categories (indicating principal function) is sufficient. This section does not apply to food which is not exposed for sale.

c) Items in (a) which have no list of ingredients and contain an ingredient which has been irradiated must be marked with a statement of that ingredient accompanied by word(s) 'irradiated' or 'treated with ionising radiation'. This section does not apply to food which is not exposed for sale.

d) Small packages and certain bottles (unless (a) or (f) apply or if (7) below applies):

   i) For any prepacked food, if either the largest surface of packaging is less than $10\,cm^2$, or, if it is contained in an indelibly marked glass bottle intended for re-use and having no label, ring or collar,

4.5

only the name and, if required, the appropriate durability indication need be given. In the case of milk, if required to avoid confusion, the place of origin is to be given, and, if raw milk, the name of manufacturer or packer. In the case of milk bottles, the appropriate durability indication need not be given.

ii) For any prepacked food which is an individual portion and is intended as a minor component to either another food or another service, only the name is required. Foods covered include butter and other fat spreads, milk, cream and cheeses, jams and marmalades, mustards, suaces, tea, coffee and sugar, and other services include provision of sleeping accommodation for trade or business.

e) Certain food sold at catering establishments:

i) Any food sold at a catering establishment which is either not prepacked or prepacked for direct sale, need not be marked with any of the items in 'General labelling requirements' above (or the additional labelling requirements (a)–(f) below).

ii) In the case of milk which is prepacked for direct sale, if required to avoid confusion, the place of origin is to be given, and, if raw milk, the name of manufacturer or packer.

iii) In the case of a food which has been irradiated, the food shall be marked with an indication of the treatment accompanied by the word(s) 'irradiated' or 'treated with ionising radiation'. In the case of irradited ingredients which would normally be declared, the ingredient shall be stated and accompanied by the appropriate word(s).

f) Seasonal selection packs: the outer packaging of a seasonal selection pack need not be marked or labelled with any particulars provided that each item in the pack meets the Regulations. A seasonal selection pack is defined as a pack consisting of two or more different items of food which are wholly or partly enclosed in outer packaging decorated with seasonal designs.

Additional requirements

a) Vending machines: when a name of a food is not visible to a purchaser, it shall be given on a notice on the front of the machine, together with (either on the machine or in close proximity):

i) for food which is not prepacked but for which a claim is made (whether on the machine or elsewhere), a notice giving the prescribed nutrition labelling;

ii) for food which should properly be reheated before it is eaten, and for which there are no reheating instructions on the label, a notice giving such instructions.

b) Alcoholic drinks: every prepacked alcoholic drink (except EC controlled wine) with greater than 1.2% alcoholic strength shall be

4.5

marked with its alcoholic strength by volume (being a figure to not more than one decimal place followed by '% vol' and which may be preceded by 'alcohol' or 'alc' and determined at 20°C). Tolerances are specified in the Regulations.

c) Raw milk: except for buffalo milk, containers of raw milk shall be marked 'This milk has not been heat-treated and may therefore contain organisms harmful to health'. In the case of any raw milk which is not prepacked and is sold at a catering establishment, the words 'Milk supplied in this establishment has not been heat-treated and may therefore contain organisms harmful to health' shall appear on a label attached to the container of milk or on a ticket or notice visible to the purchaser.

d) Products containing skimmed milk together with non-milk fat: for a product which (i) consists of skimmed milk together with non-milk fat, (ii) is capable of being used as a substitute for milk, and (iii) is neither an infant formula or a follow-on formula, nor a product specially formulated for infants or young children for medical pruposes, it shall be prominently marked with a warning that the product is unfit or not to be used for babies.

e) Foods packaged in certain gases: if the durability of a food has been extended by being packaged in a permitted packaging gas, it shall be marked 'packaged in a protective atmosphere'.

f) Foods containing sweeteners: for a food containing:
   i) a permitted sweetener, the name shall be accompanied by 'with sweetener';
   ii) both added sugar(s) and sweetener(s), the name shall be accompanied by 'with sugar(s) and sweetener(s)';
   iii) aspartame, the food shall be marked 'contains a source of phenylalanine';
   iv) more than 10% added polyols, the food shall be marked 'excessive consumption may produce laxative effects'.

Manner of marking or labelling
General:

a) When sold to the ultimate consumer, the required markings shall be either on the packaging or on a label attached to the packaging or on a label visible through the packaging. If sold otherwise than to the ultimate consumer, as an alternative, the details may be on relevant trade documents (except that the name of the food, its appropriate durability indication and the name and address of manufacturer, packer or seller must appear on the outermost packaging).

b) When those products which may omit certain details (food which is not prepacked and fancy confectionery products and certain food sold at catering establishments as detailed above in 'Omission of certain particulars' (a) and (e)) are sold to the ultimate consumer, the details which are required shall appear on a label attached to the food or on a menu, notice, ticket or label discernible to the purchaser at the place where he chooses the food. Where the information is given on a menu, etc., if the food contains (or may contain) irradiated ingredients this shall be indicated using the words 'irradiated' or 'treated with ionising radiation' accompanying the reference to the ingredient. In the case where irradiated dried substances normally used for seasoning are used in a catering establishment, an indication that food sold in the establishment contains (or may contain) those irradiated ingredients is sufficient.

c) Food which is not prepacked and fancy confectionery products (as detailed above in 'Omission of certain particulars' (a)) are sold otherwise than to the ultimate consumer, the details which are required shall appear on a label attached to the food or on a menu, notice, ticket or label discernible to the purchaser at the place where he chooses the food, or in commercial documents which accompany the food (or were sent before or at the time of delivery of the food).

Milk:
In the case of milk that is contained in a bottle, particulars may be given on the bottle cap. However, in the case of raw milk, the statement relating to health (see 'Additional requirements' (c) above) shall be given elsewhere than on the bottle cap.

Intelligibility:
Any marking or notice should be easy to understand, clearly legible and indelible and, when sold to the ultimate consumer, easily visible (although at a catering establishment where information is changed regularly, information can be given by temporary media, e.g. chalk on a blackboard.). They shall not be hidden, obscured or interrupted by written or pictorial matter.

Field of vision:
When required to be marked with one or more of the following, the required information shall appear in the same field of vision:
(a) name of food; (b) appropriate durability indication; (c) indication of alcohlic strength by volume; (d) the cautionary words in respect of raw milk (see 'Additional requirements' (c) above); (e) the warning required on products consisting of skimmed milk with non-milk fat (see 'Additional requirements' (d) above); (f) statement of net quantity.
The requirements of (b), (c) and (f) do not apply to foods falling within the section on small packages and certain bottles above (see 'Omission of certain particulars' (d)).

4.5

B Claims

The following claims in the labelling or advertising of a food are prohibited:

1) A claim that a food has tonic properties (except that the use of the word 'tonic' in the description 'Indian tonic water' or 'quinine tonic water' shall not of itself constitute this claim).

2) A claim that a food has the property of preventing, treating or curing a human disease or any reference to such a property (except that the use of the claim described below under 'Claims relating to foods for particular nutritional uses' shall not of itself constitute this claim).

The following claims in the labelling or advertising of a food are only permitted where the conditions specified in the Regulations are met (only summaries of the conditions are given here and the Regulations should be consulted for full details). When considering whether a claim is being made, a reference to a substance in an ingredients list or in any nutrition labelling shall not constitute a claim.

1) Claims relating to foods for particular nutritional uses (i.e. a claim that a food is suitable, or has been specially made, for a particular nutritional purpose which includes requirements either of people whose digestive process or metabolism is disturbed, or of people who, because of special physiological conditions, obtain special benefit from the controlled consumption of certain substances, or of infants (0–12 months) or young children (1–3 years) in good health): the food must fulfil the claim; the label must indicate the aspects of compostion or process which give the food its particular nutritional characteristics; the label must give the prescribed nutrition labelling and may have additional relevant information; when sold to the ultimate consumer, the food must be prepacked and completely enclosed by its packaging.

2) Reduced or low energy value claims (i.e. a claim that a food has a reduced or low energy value except that the presence of the words 'low calorie' for a soft drink and in accordance with the requirements below on misleading descriptions does not constitute such a claim): foods claimed to have a reduced energy value must have energy no more than three-quarters of a similar food with no such claim (unless the food is an intense sweetener either on its own or mixed with another food but still significantly sweeter than sucrose); foods claimed to have a low energy value should usually have a maximum energy of 167 kJ (40 kcal) per 100 g (or 100 ml) and per normal serving (unless the food is an intense sweetener either on its own or mixed with another food but still significantly sweeter than sucrose); in the case of an uncooked food, the claim must be in the form 'a low energy or calorie or Joule food'.

3) Protein claims (i.e. a claim that a food, other than one intended for babies or young children which satisfies the conditions for (1) above,

is a source of protein): a reasonable daily consumption of the food must contribute at least 12 g of protein; foods claimed to be a rich or excellent source of protein must have at least 20% of their energy value provided by protein and in other cases at least 12%; the label must give the prescribed nutrition labelling.

4.5

4) Vitamin/mineral claims (i.e. a claim that a food, other than one intended for babies or young children which satisfies the conditions of (1) above, is a source of vitamins/minerals; a claim is not made when a name includes the name of one or more viatamins/minerals and the food consists solely of vitamins and/or minerals and certain other substances and when mineral claims are made relating to a low or reduced level of minerals): claims may only be made with respect to vitamins or minerals in the following table (using the name in the table); where (a) the claim is not confined to named vitamins or minerals then, if the food is claimed to be a rich or excellent source of vitamins or minerals, it must contain at least one half of the recommended daily amount (RDA) of two or more of the vitamins or minerals listed in the quantity reasonably expected to be consumed in one day or, otherwise, at least one-sixth; where (b) the claim is confined to named vitamins or minerals, the conditions of (a) must apply to each named vitamin or mineral; for foods to which nutrition labelling relates, the label must carry a statement of the % RDA of any vitamin or mineral involved in the claim in a quantified serving of the food or per portion (if number of portions in pack is stated); for food supplements or waters other than natural mineral waters, the label must carry a statement of the % RDA of any vitamin or mineral involved in the claim in either a quantified serving of the food or, if prepacked, per portion, and, if prepacked, the number of portions in the pack is to be stated:

	Recommended daily allowance
Vitamins	
Vitamin A	800 μg
Vitamin D	5 μg
Vitamin E	10 mg
Vitamin C	60 mg
Thiamin	1.4 mg
Riboflavin	0.16 mg
Niacin	18 mg
Vitamin B_6	2 mg
Folacin	200 μg
Vitamin B_{12}	1 μg
Biotin	0.15 mg
Pantothenic acid	6 mg
Minerals	
Calcium	800 mg
Phosphorus	800 mg

4.5

(*continued*)

	Recommended daily allowance
Iron	14 mg
Magnesium	300 mg
Zinc	15 mg
Iodine	150 µg

Note: For the purposes of prescribed nutrition labelling, a 'significant amount' means 15% of the RDA listed that is supplied by 100 g or 100 ml of a food or per package if the package contains only one portion.

5) Cholesterol claims (i.e. a claim relating to the presence or absence of cholesterol): the food must have a maximum of 0.005% cholesterol except that if it is higher than this figure a claim may be made if the claim relates to the removal of cholesterol from, or its reduction in the food if the claim is made (a) as part of an indication of the true nature of the the the food, (b) as part of an indication of the treatment of the food, (c) within the list of ingredients, or (d) as a footnote in respect of prescribed nutrition labelling; the claim must not include any suggestion of benefit to health because of its level of cholesterol; the food shall be marked with the prescribed nutrition labelling.

6) Nutrition claims (i.e. a claim not dealt with under items (1)–(5) above): the food must be capable of fulfilling the claim; the food shall be marked with the prescribed nutrition labelling.

7) Claims which depend on another food (i.e. a claim that a food has a particular value or confers a particular benefit): the value or benefit must not be derived wholly or partly from another food intended to be consumed with the food.

C Prescribed nutrition labelling

Presentation

It shall be presented together in one conspicuous place in tabular form with numbers aligned or, if there is insufficient space for this, in linear form. When required or permitted to be given, the following order and manner of listing shall be used:

	(1)
Energy	[x] kJ and [x] kcal
Protein	[x] g
Carbohydrate, of which:	[x] g
sugars	[x] g
polyols	[x] g
starch	[x] g
Fat, of which:	[x] g
saturates	[x] g

(continued)

	(1)
monounsaturates	[x] g
polyunsaturates	[x] g
cholesterol	[x] mg
Fibre	[x] g
Sodium	[x] g
[Vitamins] (2)	[x units] (3)
[Minerals] (2)	[x units] (3)

Notes:
  (1) [x] to be substituted by the appropriate amount
  (2) The name(s) to be given as used in the table of vitamins and minerals given above
  (3) [x units] to be substituted by the amount using the units in the table of vitamins and minerals given above, and to give the % RDA

Where it is required to give additional information relating to any substance which belongs to, or is a component of, one of the items listed, it shall appear as follows:

[item] (1), of which:	[x] g or mg
[substance or component] (2)	[x] g or mg

Notes:
  (1) [item] to be the relevant item from above list
  (2) [substance or component] to be the relevant name

All amounts:
  a) to be given per 100 g or 100 ml;
  b) may, in addition, be given per quantified serving or per portion (if number of portions in pack is stated);
  c) to be the amount contained in the food as sold except that, where detailed preparation instructions are given, they may be the amounts after such preaparation (and this must be expressly indicated);
  d) to be averages (taking into account seasonal variation, patterns of consumption, etc.) based on (i) the manufacturer's analysis of the food, (ii) a calculation from the actual average values of the ingredients used in the preparation of the food, and/or (iii) a calculation from generally established and accepted data.

In calculating the energy value the following conversion factors shall be used:

1 g carbohydrate (excluding polyols) = 17 kJ (4 kcal)
1 g polyols = 10 kJ (2.4 kcal)
1 g protein = 17 kJ (4 kcal)
1 g fat = 37 kJ (9 kcal)
1 g ethanol = 29 kJ (7 kcal)
1 g organic acid = 13 kJ (3 kcal)

4.5

Contents

1) Except where (2) applies, prescribed nutrition labelling shall give the following:

   a) it shall include either:

      i) energy, protein, carbohydrates and fat, or

      ii) energy, protein, carbohydrates, sugars, fat, saturates, fibre and sodium;

   b) if a claim is made for any of sugars, saturates, fibre or sodium, then it shall be given according to (a) (ii);

   c) where a nutrition claim is made relating to polyols, starch, monounsaturates, polyunsaturates, cholesterol, vitamins or minerals, the relevant amount shall be given except that, in the case of vitamins or minerals, the amount present must be a significant amount (15% of the RDA – see table on page 171);

   d) the items in (c) may be given even if no claim is made but the restriction relating to vitamins or minerals also applies;

   e) if the labelling is presented in the form (a) (i) above, but includes monounsaturates, polyunsaturates, and/or cholesterol, the amount of saturates must also be given;

   f) where a nutrition claim is made relating to any substance which belongs to, or is a component of, one of the nutrients already required (or permitted) to be included, the name and amount of that substance shall be given.

2) For food which is not prepacked and which is either sold to the ultimate consumer other than at a catering establishment, to the ultimate consumer from a vending machine, or to a catering establishment, prescribed nutrition labelling shall give any data relevant to any nutrtiion claim which is made and may include that shown above under 'Presentation'.

D Misleading descriptions

1) General: the following words and descriptions may only be used in the labelling or advertising of the foods indicated in the following table if the foods satisfy the conditions stated:

Words and descriptions	Conditions
'Dietary' or 'dietetic'	Only to be used for foods for a particular nutritional use (excluding foods for infants and young children in good health) if: a) it has been specially made for a class of person whose digestive process or metabolism is disturbed or who, because of a special physiological condition, obtains special benefit from a controlled consumption of certain substances, and b) it fulfils those particular nutritional requirements.

(continued)

4.5

Words and descriptions	Conditions
Food with an implied flavour	a) Flavour to be derived wholly or mainly from the food named in the description, except 'chocolate' may be used when the flavour is derived wholly or mainly from non-fat cocoa solids (and the purchaser would not be misled); b) the word 'flavour' may be used preceded by the name of a food if (a) does not apply.
Pictorial representation to imply a flavour	Flavour to be derived wholly or mainly from the food depicted in the representation
'Ice-cream'	Food must be a frozen product containing min. 5% fat and min. 2.5% milk protein and which is obtained by subjecting an emulsion of fat, milk solids and sugar (any any permitted sweetener) with or without the addition of other substances, to heat treatment and either to subsequent freezing or evaporation, addition of water and subsequent freezing.
'Dairy ice-cream'	Must meet the conditions for 'ice cream' above (provided that the min. 5% fat shall consist exclusively of milk fat) and it contains no fat other than milk fat or any fat present by reason of the use as an ingredient of any egg, flavouring, emulsifier or stabiliser
'Milk' or any other word or description with implied milk content	Only to be used as part of the name of a food, if it contains cows' milk unless: a) it comes from another named animal and it has all the normal constituents in their natural proportions; b) it comes from another named animal and has been subject to a named process or treatment; or c) it meets the requirements of other regulations.
'Milk'	Only to be used as part of the name of an ingredient if it is cows' milk unless it is accompanied by the name of another animal and its use complies with other aspects of the regulations.
'Starch-reduced'	Only to be used if less than 50% (weight, dry matter) of food consists of anhydrous carbohydrate and starch content is substantially less than similar foods not so described.
'Vitamin' or similar description implying vitamin	Only to be used if the food described is one of the vitamins in the above table of vitamins and minerals, or vitamin K
'Alcohol-free'	Only to be used for an alcoholic drink from which the alcohol has been extracted if the alcoholic strength is max. 0.05% (vol.) and the drink is marked with either the maximum value as % vol. preceded by 'not more than' or an indication that it contains no alcohol.

4.5

(continued)

Words and descriptions	Conditions
'Dealcoholised'	Only to be used for an alcoholic drink from which the alcohol has been extracted if the alcoholic strength is max. 0.5% (vol.) and the drink is marked with either the maximum value as % vol. preceded by 'not more than' or an indication that it contains no alcohol.
'Low alcohol' or similar description implying the drink is low in alcohol	Only to be used for an alcoholic drink if the alcoholic strength is max. 1.2% (vol.) and the drink is marked with the maximum value as % vol. preceded by 'not more than'.
'Low calorie' or similar description implying drink is low in calories	Only to be used for a soft drink if it (or if it after any specified preparation) contains max. 42 kJ (10 kcal) per 100 ml.
'Non-alcoholic'	Not to be used with a name commonly associated with an alcoholic drink except in the form 'non-alcoholic wine' described in the section on wine below.
'Liqueur'	Not to be applied to any drink other than one meeting the definition of liqueur contained in EC Regulations.
'Indian tonic water' or 'quinnine tonic water'	Only to be used for a drink containing min. 57 mg quinnine per litre.
'Tonic wine'	Only to be used for a drink if in immediate proximity to the words 'tonic wine' there is the statement: 'the name 'tonic wine' does not imply health giving or medicinal properties', and no recommendation as to consumption or dosage shall be given.

2) Cheese: the following names may not be used in the labelling of any cheese unless the cheese has at least 48% milk fat (dry matter basis) and the cheese has no more than the following specified water contents:

Variety of cheese	Maximum % of water
Cheddar	39
Blue Stilton	42
Derby	42
Leicester	42
Cheshire	44
Dunlop	44
Gloucester	44
Double Gloucester	44
Caerphilly	46
Wensleydale	46
White Stilton	46
Lancashire	48

3) Cream: the following names may not be used in the labelling of any cream unless the specified requirements are met, except that the milk

fat requirement need not be met if the name contains qualifying words indicating that the milk fat content of the cream is greater or less than that specified:

Name of cream	Minimum % milk fat	Other requirements
Clotted cream	55	The cream is clotted
Double cream	48	
Whipping cream	35	
Whipped cream	35	The cream has been whipped
Sterilised cream	23	The cream is sterilised cream
Cream or single cream	18	The cream is not sterilised cream
Sterilised half cream	12	The cream is sterilised cream
Half cream	12	The cream is not sterilised cream

4) 'Wine' used in a composite name: the use of the word 'wine' is restricted by EC Regulations. However, it may be used in a composite name in the following cases (so long as no confusion is caused):
   a) 'Non-alcoholic wine' may only be used for a drink derived from unfermented grape juice intended exclusively for communion or sacramental use (and labelled as such);
   b) When the word 'wine' is used in a composite name for a drink which is derived from fruit other than grapes, that drink shall be obtained by an alcoholic fermentation of that fruit.

Milk and milk products

Regulations: Milk and Milk Products (Protection of Designations) Regulation 1990 (1990/607)
   EC Council Regulation 1898/87
   EC Commission Decision 88/566
Amendments: Food (Miscellaneous Revocations and Amendments) Regulations 1995 (1995/3267)
   Food Labelling Regulations 1996 (1996/1499)

Definitions
Milk: the normal mammary secretion obtained from one or more milkings without either addition thereto or extraction therefrom.
Milk products: products derived exclusively from milk, on the understanding that substances necessary for their manufacture may be added provided that those substances are not used for the purpose of replacing, in whole or in part, any milk constituent.

General points
   1) Subject to (2) below, the EC Regulation reserves exclusively the following:
      a) for milk, the designation 'milk' except that it may be used for

4.5

standardised milk and with other words indicating a process or treatment which has added or withdrawn natural constituents;
   b) for milk products, the following designations – whey, cream, butter, buttermilk, butteroil, caseins, anhydrous milk fat (AMF), cheese, yoghurt, kephir, koumiss, and designations or names actually used for milk products and used in accordance with the 'name' requirements of the Food Labelling Regulations.
2) The designations given in (1) may be used on products, contrary to (1), where the exact nature of the product is clear from traditional usage and/or when the designations are clearly used to describe a characteristic quality of the product. An indicative list has been published (EC Commission Decision 88/566) which includes the following UK products which are considered to meet this exception: coconut milk; 'cream ...' or 'milk' used in the description of a spiritous beverage not containing milk or other milk products or milk or milk product imitations (e.g. cream sherry, milk sherry); cream soda; cream filled biscuits (e.g. custard cream, bourbon cream, raspberry cream biscuits, strawberry cream, etc.); cream crackers; salad cream; creamed coconut and other similar fruit, nut and vegetable products where the term 'creamed' describes the characteristic texture of the product; cream of tartar; cream or creamed soups (e.g. cream of tomato soup, cream of celery, cream of chicken, etc.); horseradish cream; ice-cream; jelly cream; table cream; cocoa butter; shea butter; nut butters (e.g. peanut butter); butter beans; butter puffs; fruit cheese (e.g. lemon cheese, damson cheese).

Lot marking

Regulation: Food (Lot Marking) Regulations 1996 (1996/1502)

Definitions
Lot: a batch of sales units of food produced, manufactured or packaged under similar conditions.
Lot marking indication: an indication which allows identification of the lot to which a sales unit of food belongs.

Lot marking requirements
1) Food which forms part of a lot must be accompanied by a lot marking indication.
2) For the purposes of (1):
   a) a lot shall be determined as a lot to which food in the sales unit belongs by a producer, manufacturer, packager, or the first seller established within the EC, of the food in question;
   b) a lot marking indication will be determined and fixed by one of the operators in (a) and shall be preceded by the letter 'L' unless it

is clearly distinguishable from other indications on the packaging or label.

Exemptions

The following are exempt from the requirements:

a) agricultural products when sold to a temporary storage, preparation or packaging station or to a producers' organisation or for collection for immediate integration into an operational preparation or processing system;

b) food which is sold to the ultimate consumer and is not prepacked, is packed at the request of the purchaser or is prepacked for immediate sale;

c) sales units in containers where the area of the largest side is less than $10 \, cm^2$;

d) a sales unit which is prepacked, sold as an individual portion for immediate consumption, and is intended as a minor accompaniment to another food or service;

e) a sales unit of an individual portion of an edible ice supplied to its seller in bulk packaging which bears the required lot marking indication;

f) a sales unit marked or labelled before 1 July 1992 and until 1 July 1997, re-usable indelibly marked glass bottles with no label, ring or collar;

g) a sales unit which is marked or labelled with an indication of minimum durability or 'use by' date consisting of at least the uncoded indication of the day and month (as required by the Food Labelling Regulations, or in accordance with their requirements).

Food additives

Regulation: Food Additives Labelling Regulations 1992 (1992/1978)
Amendments: Sweeteners in Food Regulations 1995 (1995/3123)
  Colours in Food Regulations 1995 (1995/3124)
  Miscellaneous Food Additives Regulations 1995 (1995/3187)
  Food Labelling Regulations 1996 (1996/1499)

Definitions

Food additive: such food as comprises material (with or without nutritive value): (a) which is within one of the categories in the following table; (b) which is neither normally consumed as a food in itself nor normally used as a characteristic ingredient of food; (c) which is not a processing aid (see definition), an approved plant or plant product protection substance, a permitted flavouring (see Flavouring Regulations) or a substance added to food as a nutrient; and (d) the intentional addition of which to other food for a technological purpose in the manufacture, processing, preparation, treatment, packaging,

4.5

transport or storage of that other food results, or may be reasonably expected to result, in that material or its byproducts becoming directly or indirectly a component of that other food.

Processing aid: substances: (a) which are not consumed as food ingredients by themselves; (b) which are intentionally used in the processing of raw materials, foods or their ingredients, to fulfil technological purposes during treatment or processing into finished products; (c) which are capable of resulting in the unintended but technically unavoidable presence of residues of such substances (or their derivatives) in the finished products; and (d) the residues of which (or, as the case may be, of the derivatives of which) do not present any risk to human health and do not have any technological effect on the finished products.

Supplementary material: material (whether or not it comes within the excluded materials mentioned in item (c) of the definition of food additive above) the presence of which in or on the food additive may reasonably be expected to facilitate storage, sale, standardisation, dilution or dissolution of the food additive.

The following are the categories of additives specified in the definition of 'food additives':

Colour	Acids	Flour treatment agents
Antioxidants	Acidity regulators	Firming agents
Preservatives	Anti-caking agents	Humectants
Emulsifiers	Modified starch	Enzyme preparations
Emulsifying salts	Sweeteners	Sequestrants
Thickeners	Raising agents	Bulking agents
Gelling agents	Anti-foaming agents	Propellants
Stabilisers	Glazing agents	Packaging gas
Flavour enhancers	Flour bleaching agents	Carriers and carrier solvents

Labelling requirements

1) The following details shall be given:
    a) the name of the product being either the legally specified additive name(s) and E number(s), or, where there are no legal names or numbers, a description sufficiently precise to enable them to be distinguished from any product(s) with which they could be confused (given in weight descending order);
    b) where supplementary material is present, an indication of each component of it in weight descending order;
    c) the statement 'for use in food' or 'restricted use in food' or a more specific use statement;
    d) any special storage conditions;
    e) any special conditions of use;

f) any instructions for use if it would be difficult to make appropriate use of the food additives in the absence of such instructions;

4.5

g) a batch or lot identifying mark;

h) the name and address of the manufacturer, packer or seller (in the EEC);

i) for business sales only, where an additive has legally restricted use in food, the proportion of the restricted additive in the material sold or information allowing its legal use;

j) for consumer sales only, an indication of minimum durability (as specified by the Food Labelling Regulations).

2) For business sales, the information given may be:

a) as (a) to (i) in (1), or

b) the container should be labelled with items (a), (c), (d) and (e), shall bear on the label a conspicuous statement 'intended for manufacture of foodstuffs and not for retail sale', and the remaining items in (1) may appear on trade documents.

General point

The Regulations do not apply to any additive which has become part of other food (i.e. when it is added to other food which comprises or contains material other than food additives except in the case of the addition of supplementary material).

Organic foods

Regulations: Organic Products Regulations 1992 (1992/2111)
   EC Council Regulation 2092/91
Amendments: EC Commission Regulations 1535/92, 2083/92, 3713/92, 207/93, 1593/93, 2608/93, 468/94, 688/94, 1468/94, 2381/94, 2580/94, 1201/95, 1202/95, 1935/95, 418/96
   Organic Products (Amendment) Regulations 1993 (1993/405)
   Organic Products (Amendment) Regulations 1994 (1994/2286)

Definition

Organic products: the following products, where such products bear, or are intended to bear indications referring to organic production methods: (a) unprocessed agricultural crop products and animals and unprocessed animal products (to the extent that these are covered by the EC Regulations); (b) products intended for human consumption composed essentially of one or more ingredients of plant origin or, to the extent that they are covered, ingredients of animal origin.

General points

1) The word 'organic' may only be applied to products which have been obtained in accordance with the requirements of the EC Regulation

4.5

2092/91 (unless not applied to agricultural products in foodstuffs or clearly has no connection with the method of production).

2) The labelling and advertising of products in part (a) of the above definition may refer to organic production methods only where: the reference clearly relates to a method of agricultural production; the product is produced in the EC (or imported from a third country) in accordance with the requirements of the EC Regulation 2092/91; the producer (or importer) has been subject to specified inspection; for products prepared after 1 January 1997, the labelling refers to the name and/or code number of the inspection authority.

3) The labelling and advertising of products in part (b) of the above definition may refer to organic production methods in the sales description only where: at least 95% of the ingredients of agricultural origin are obtained in accordance with the specified procedures; all the other ingredients of agricultural origin are those permitted by the Regulation or provisionally authorised by a Member State; the product contains only certain specified substances of non-agricultural origin; the products or its ingredients have only been treated with certain substances; the product (or ingredients) have not been subject to irradiation; and the preparer (or importer) is subject to specified inspection; for products prepared after 1 January 1997, the labelling refers to the name and/or code number of the inspection authority. The reference to organic production methods must make it clear that they relate to a method of agricultural production and indicate the ingredients concerned (unless it is clear from the ingredients' list).

4) Products consisting of one ingredient of agricultural origin which do not meet the full conversion period required for organic production may be labelled or advertised in accordance with (2) or (3) provided that they meet the other requirements of (2) and (3) and a conversion period of at least 12 months has been met and the indication takes the form 'product under conversion to organic farming'.

5) The labelling and advertising of products in part (b) of the above definition may refer to organic production methods only where: at least 70% of the ingredients of agricultural origin meet the requirements; the product meets the other requirement of (3) above; the reference is in the ingredients' list (in the same colour, size and style as other words and clearly refers to the organic ingredients) and there may be a statement 'x% of the agricultural ingredients were produced in accordance with the rules of organic production' in the same field of vision as the sales description.

6) Until 31 December 1997, for products in part (b) of the above definition prepared partly from ingredients of agricultural origin which are not obtained in accordance with the specified procedures, reference to organic production methods may only be used on products if: at least 50% of the ingredients meet the specified

requirements; the product meets the other requirements of (3) above; the reference is only in the ingredients' list (in the same colour, size and style as other words and clearly refers to the organic ingredients).

7) Where a product is covered by an inspection scheme and meets certain other criteria (see Regulations) the indication 'Organic Farming – EEC Control System' may be used. No claim may be made to suggest that this indication constitutes a guarantee of superior organoleptic, nutritional or salubrious quality.
8) For the UK, Ministers have designated responsibility for inspection and the scheme is operated by 'Food from Britain'.

4.6 Hygiene and health

General

Regulation: Food Safety (General Food Hygiene) Regulations 1995 (1995/1763)
Amendment: Dairy Products (Hygiene)(Amendment) Regulations 1996 (1996/1699)

Note: The Regulations do not apply to primary production nor to those activities regulated by other Regulations on hygiene (and summarised on pages 190–217) except that:
   a) all food businesses must have a supply of, and must use, potable water (unless other Regulations specify alternative requirements or if on board a fishing vessel) (see Chapter VII (1), below);
   b) all food businesses must meet the training requirements (unless other Regulations specify alternative requirements) (see Chapter X, below);
   c) the 'General requirements' and 'Exclusion for medical conditions' sections apply to catering establishments and shops handling or selling any heat-treated cream or any ice-cream (as defined in the Dairy Products (Hygiene) Regulations 1995).

Definitions
Hygiene: all measures necessary to ensure the safety and wholesomeness of food during preparation, processing, manufacturing, packaging, storing, transportation, distribution, handling and offering for sale or supply to the consumer (and 'hygienic' shall be construed accordingly).
Primary production: includes harvesting, slaughter and milking.

General Requirements
   1) Food businesses must carry out their operations (i.e. those mentioned in the definition of hygiene) in a hygienic way.
   2) Food businesses shall identify any step in their activities which is

4.6 critical to ensuring food safety and ensure that adequate safety procedures are identified, implemented, maintained and reviewed on the basis of the following principles:

a) analysis of the potential food hazards in a food business operation;

b) identification of the points in those operations where food hazards may occur;

c) deciding which of the points identified are critical to ensuring food safety ('critical points');

d) identification and implementation of effective control and monitoring at those critical points; and

e) review of the analysis of food hazards, the critical points and the control and monitoring procedures periodically, and whenever the food business' operations change.

Exclusion for medical conditions

Any person who either knows/suspects that they are a carrier of a disease likely to be transmitted through food, or is afflicted with an infected wound, a skin infection, sores, diarrhoea (or any analogous medical condition), and is likely to directly or indirectly contaminate food with pathogenic micro-organisms shall inform the proprietor of the food business at which they are working. This requirement only applies to those people who work in situations where a proprietor may be required to exclude them under the specific provisions relating to personal hygiene (see Chapter VIII, below).

Specific requirements

The Regulations specify in a Schedule specific 'Rules of Hygiene'. They should be consulted for full details – the following is a summary of the main points.

Chapter I – General requirements for food premises (other that those specified in Chapter III)

1) Food premises must be kept clean and maintained in good repair and condition.

2) The layout, design, construction and size of food premises shall

a) permit adequate cleaning and/or disinfection,

b) protect against the accumulation of dirt, contact with toxic materials, the shedding of particles into food and the formation of condensation or undesirable mould on surfaces,

c) permit good hygiene practices (including protection from cross-contamination and contamination from external sources such as pests),

d) provide suitable temperature conditions for the hygienic processing and storage of products.

3) An adequate number of washbasins (with hot/cold/mixed running water, materials for cleaning and hygienic drying) and lavatories (which must not lead directly into food handling rooms, and must be provided with adequate natural or mechanical ventilation) must be available. Where necessary, food washing provision must be separate from hand-washing facility.

4) Suitable and sufficient natural or mechanical ventilation must be provided.

5) Food premises must have adequate natural and/or artificial lighting.

6) Drainage facilities must be adequate for the purpose intended and must be designed and constructed to avoid the risk of contamination of foodstuffs.

4.6

Chapter II – Specific requirements in rooms where foodstuffs are prepared, treated or processed (excluding dining areas and those premises specified in Chapter III)

1) In rooms where food is prepared, treated or processed:
   a) floor and wall surfaces and surfaces in contact with food must be maintained in a sound condition, and must be easy to clean and, where necessary, disinfect;
   b) doors must be easy to clean and, where necessary, disinfect;
   c) ceilings and overhead fixtures must be designed, constructed and finished to prevent accumulation of dirt and reduce condensation, the growth of undesirable moulds and the shedding of particles;
   d) windows and other openings must either be constructed to prevent the accumulation of dirt and, where necessary, be fitted with insect-proof screens (easily removed for cleaning) or must remain closed if contamination of foodstuffs would result if they are opened.

2) Where necessary, adequate facilities (corrosion resistant and supplied with adequate hot and cold water) must be provided for cleaning and disinfecting of work tools and equipment.

3) Where appropriate, adequate provision must be made for any necessary washing of the food (sinks or other facility to be adequately supplied with hot and/or cold potable water and be kept clean).

Chapter III – Requirements for movable and/or temporary premises, premises used primarily as a private dwelling house, premises used occasionally for catering purposes and vending machines

1) Premises and vending machines shall be so sited, designed, constructed, and kept clean and maintained in good repair and condition, as to avoid the risk of contaminating foodstuffs and harbouring pests, so far as reasonably practicable.

2) In particular,
   a) adequate provision should be made for: personal hygiene;

4.6

cleaning (and, if necessary, disinfecting) work utensils and equipment; cleaning of foodstuffs; hot and/or cold potable water; hygienic storage and disposal of hazardous/inedible substances and waste; maintaining and monitoring suitable food temperature conditions;

b) surfaces in contact with food must be maintained in a sound condition, and must be easy to clean and, where necessary, disinfect;

c) foodstuffs must be so placed as to avoid, so far as reasonably practicable, the risk of contamination.

Chapter IV – Transport

1) Conveyances and/or containers used for transporting foodstuffs must be kept clean and maintained in good repair and condition in order to protect foodstuffs from contamination, and must, where necessary, be designed and constructed to permit adequate cleaning and/or disinfection.

2) Receptacles in vehicles and/or containers must not be used for transporting anything other than foodstuffs where this may result in contamination of foodstuffs.

3) Bulk foodstuffs in liquid, granular or powder form must be transported in receptacles and/or containers/tankers reserved for the transport of foodstuffs (and marked in a specified way) if otherwise there is a risk of contamination.

4) If food is transported with other foods or anything else, there must be effective separation of products to protect against the risk of contamination, and there must be effective cleaning between loads.

5) Conveyances and/or containers must be capable of maintaining appropriate temperatures and, where necessary, designed to allow those temperatures to be monitored.

Chapter V – Equipment requirements

All article, fittings and equipment with which food comes into contact shall be kept clean and:

a) be so constructed, be of such materials and be kept in such good order, repair and condition, as to both minimise any risk of contamination of the food and (with the exception of non-returnable containers and packaging) to enable them to be kept thoroughly cleaned (and, if necessary, disinfected) sufficient for the intended purpose;

b) be installed in such a manner as to allow adequate cleaning of the surrounding area.

Chapter VI – Food waste

1) Food waste and other refuse must:

a) not be allowed to accumulate in food rooms, except so far as is unavoidable for the proper functioning of the business,

b) be deposited in closable containers (unless other containers are appropriate) of a suitable design;

2) There must be adequate provision for the removal and storage of food waste and other refuse. Refuse stores must be able to be kept clean and protected against pests and against contamination of food, drinking water, equipment or premises.

4.6

Chapter VII – Water supply

1) There must be an adequate supply of potable water.

2) Where appropriate, ice must be made from potable water and must be made, handled and stored under conditions which protect it from contamination.

3) Steam used directly in contact with food must not contain any substance which presents a hazard to health, or is likely to contaminate the product.

4) Water unfit for drinking (used for non-food purposes) must be conducted in readily identifiable separate systems having no connection with the potable water systems.

Chapter VIII – Personal hygiene

1) Every person working in a food handling area shall maintain a high degree of personal cleanliness and shall wear suitable, clean and, where appropriate, protective clothing.

2) No person known or suspected to be suffering from, or to be a carrier of, a disease likely to be transmitted through food or while afflicted shall be permitted to work in any food handling area in any capacity in which there is any likelihood of directly or indirectly contaminating food with pathogenic micro-organisms.

Chapter IX – Provisions applicable to foodstuffs

1) No raw materials or ingredients shall be accepted by a food business if they are (or expected to be) contaminated with parasites, pathogenic micro-organisms, or toxic, decomposed or foreign substances that, after normal sorting and/or hygienic processing, would still be unfit for human consumption.

2) Storage of raw materials and ingredients must prevent their harmful deterioration and protect them from contamination.

3) All food must be protected against contamination likely to render the food unfit, injurious to health or unreasonably contaminated. Procedures must ensure that pests are controlled.

4) Hazardous and/or inedible substances (including animal feedstuffs) shall be adequately labelled and stored in separate and secure containers.

4.6

Chapter X – Training

The proprietor of a food business shall ensure that food handlers engaged in the food business are supervised and instructed and/or trained in food hygiene matters commensurate with their work activities.

General points

In enforcing the Regulations, food enforcement authorities shall:

a) ensure that premises are inspected with a frequency which has regard to the risk associated with those premises;

b) ensure that inspections include a general assessment of the potential food safety hazards associated with the food business being inspected;

c) assess the operation of the monitoring and verification controls for identified critical control points;

d) give consideration to whether the business has acted in accordance with any recognised guide to good hygiene practice (See Appendix 3 for listing of recognised guides).

Temperature controls

Regulation: Food Safety (Temperature Control) Regulations 1995 (1995/2200)
Amendment: Food Labelling Regulations 1996 (1996/1499)

Notes:

a) These Regulations contain different provisions for England and Wales and for Scotland. The details given below are based on the requirements for England and Wales.

b) The Regulations do not apply to primary production or to those activities regulated by other Regulations on hygiene (and summarised on pages 190–217) except for certain aspects relating to fishery products where they do apply unless other requirements are specified in the relevant fishery product regulations.

Definition

Shelf-life: For foods marked with either a 'best before' date or a 'use by' date, the period up to and including that date; for foods not required to be marked, the period for which the food can be expected to remain fit for sale if it is kept in a manner which is consistent with food safety.

General requirement

No person, in the course of a food business, shall keep foodstuffs which are:

a) raw materials, ingredients, intermediate products or finished products; and

b) likely to support the growth of pathogenic micro-organisms or the

4.6

formation of toxins, at temperatures which would result in a risk to health. This may require foods to be held at specified temperatures below 8°C. However, consistent with food safety, limited periods outside temperature control are permitted where necessary (i.e. for practicalities of handling during preparation, transport, storage, display and service of food).

Chill holding requirements

1) Subject to (2)–(4) below, no person carrying out commercial operations, shall keep any food which is likely to support the growth of pathogenic micro-organisms or the formation of toxins at or in food premises at a temperature above 8°C.

2) Where food which, as part of a mail order transaction, is being conveyed by post to an ultimate consumer, (1) does not apply. However, no food shall be supplied by mail order which is being conveyed by post to an ultimate consumer, which is likely to support the growth of pathogenic micro-organisms or the formation of toxin at a temperature which has given rise to or is likely to give rise to a risk to health.

3) The following are exempt from the requirements of (1) and (2) above:
   a) foods required to be held hot by the 'Hot holding requirement' (see below);
   b) foods which, during their shelf-life, may be kept at ambient temperatures with no risk to health;
   c) foods which are being (or have been) subject to a process (e.g. by dehydration or canning) intended to prevent the growth of pathogenic micro-organisms at ambient temperatures. In the case of canned foods (or similar), once the container is opened, this exemption no longer applies;
   d) foods which require ambient temperatures for ripening or maturing (but only for the duration of the ripening or maturing);
   e) raw food intended for processing if the process will render the food fit for consumption;
   f) food subject to certain EC Regulations on marketing of poultry (Reg. 1906/90) and of eggs (Reg. 1907/90).

4) If a food manufacturer, preparer or processor has recommended a storage temperature for a food between 8°C and ambient for a specified shelf-life (based on a well-founded scientific assessment of the safety of the food at the recommended temperature), then the food can be kept in compliance with the recommendations.

5) Foods which are required by the above points to be kept below ambient temperature shall be cooled as quickly as possible following either the final heat processing stage or, if no heat process is applied, the final preparation stage.

189

4.6 *Chill holding tolerance periods*

1) Food for service or on display for sale may be kept above 8°C for a period of up to 4 hours (if it had not previously been kept, for service or on display for sale, at above 8°C).

2) Food being transferred from (or to) a vehicle used by the food business to (or from) premises at which the food was going to be kept at 8°C may be kept above 8°C for a limited period only (consistent with food safety).

3) If there is an unavoidable reason (e.g. practicalities of handling during processing, defrosting of equipment, or temporary breakdown of equipment), food may be kept above 8°C for a limited period only (consistent with food safety).

Hot holding requirements

1) Except as specified in (2) and (3), no person carrying out commercial operations shall keep food which has been cooked or reheated, is for service or on display for sale and which needs to be kept hot in order to control the growth of pathogenic micro-organisms or the formation of toxins, at a temperature below 63°C.

2) If there is no risk to health (based on a well-founded scientific assessment of the safety of the food at the recommended temperature) when foods, covered by (1), are kept below 63°C for a specified time, then the requirements of (1) do not apply.

3) Food for service or on display for sale may be kept below 63°C for a period of up to 2 hours (if it had not previously been kept, for service or on display for sale).

Milk and milk-based products

A Infection of milk

Regulation: Milk and Dairies (General) Regulations 1959 (1959/277)
Amendments: Milk and Dairies (Standardisation and Importation) Regulations 1992 (1992/3143)
Dairy Products (Hygiene) Regulations 1995 (1995/1086)

Definition

Milk: Cows' milk intended for sale or sold for human consumption or intended for manufacture into products for sale for human consumption, and includes cream and skimmed milk, as well as standardised whole milk and non-standardised whole milk.

Requirements

Following the adoption of the Dairy Products (Hygiene) Regulations, these Regulations now contain only residual controls relating to the infection of milk:

1) Where a person who has access to milk (or milk containers) becomes aware that they or a member of their household is suffering from a notifiable disease they shall inform the occupier of the registered premises who shall inform the medical officer of health for the district.

2) Medical officers of health who suspect that someone may be suffering from a disease liable to cause infection of milk may issue a notice preventing the person from working with milk.

3) Medical officers of health who are satisfied (or have reasonable grounds to suspect) that someone is suffering from disease caused by the consumption of milk supplied from a registered premises may restrict the sale of the milk from the premises.

4.6

B Dairy products

Regulation: Dairy Products (Hygiene) Regulations 1995 (1995/1086)
Amendments: Dairy Products (Hygiene)(Amendment) Regulations 1996 (1996/1699)
  Food Labelling Regulations 1996 (1996/1499)

Definitions

Dairy product: milk or any milk-based product.

Milk: the milk of cows, ewes, goats or buffaloes intended for human consumption.

Milk-based product: (a) a milk product exclusively derived from milk to which other substances necessary for its manufacture may have been added, provided that those substances do not replace in part or in whole any milk constituent, and (b) a composite milk product of which no part replaces or is intended to replace any milk constituent and of which milk or a milk product is an essential part either in terms of quantity or for characterisation of the product.

Dairy establishment: any undertaking handling dairy products and is either (a) a standardisation centre, or any one of the following undertakings operating alone or in combination: (b) a treatment establishment, (c) a processing establishment, or (d) a collection centre.

Processing establishment: an establishment where any diary product is either treated, processed and wrapped or undergoes one or more of those handling activities.

Standardisation centre: an establishment which is not attached to a collection centre or to a treatment or processing establishment and where raw milk may be skimmed or its natural constituents modified.

Treatment establishment: an establishment where milk is heat-treated.

Collection centre: an establishment where raw milk is collected and where it may be cooled and filtered.

Production holding: premises at which one or more milk-producing cows, ewes, goats or buffaloes are kept.

4.6 *Main requirements*

1) Only registered production holdings complying with detailed requirements (relating to (a) hygiene, (b) housing of animals, (c) the milking, handling, cooling and storage of raw milk, and (d) milking and filtering operations) may be used for the production of raw milk. Registration is performed by the Ministry. Only milk which complies with the requirements of the Regulations may be dispatched from a production holding.

2) Only approved dairy establishments complying with detailed requirements (relating to (a) general conditions of hygiene for dairy establishments, (b) special conditions for approval of treatment or processing establishments, and (c) general conditions of hygiene applicable to dairy establishments, instruments and equipment) may be used. Approval is provided by a food authority. Only dairy products which comply with the requirements of the Regulations may be dispatched from a dairy establishment.

3) The registration of a production holding may be cancelled and the approval of a dairy establishment may be revoked under specified conditions.

4) Only raw milk, thermised milk, heat-treated drinking milk, heat-treated milk intended for the manufacture of milk-based products, or milk-based products meeting the detailed requirements of the Regulations may be sold for human consumption. Some of the requirements are summarised below (see Regulations for full details).

5) Any cows' milk sold to a catering establishment must be heat-treated.

6) No thermised cows' milk may be sold to the ultimate consumer without heat-treatment.

7) No ice-cream (which is a milk-based product) may be sold unless it has been pasteurised or sterilised.

8) In the case of cheese with a period of ageing or ripening of at least 60 days and which meets the microbiological criteria below, certain exemptions apply including: (a) reduced hygiene requirements, (b) reduced criteria for the raw milk used, and (c) exemption from the wrapping and packaging requirements

9) In the case of milk-based products with traditional characteristics certain exemptions apply including: (a) reduced hygiene requirements, (b) reduced criteria for the raw milk used, (c) exemption from the microbiological criteria, and (d) exemption from the wrapping and packaging requirements.

10) Dairy products intended for sale must: (a) be handled, stored and transported under specified conditions, (b) be wrapped and packaged in specified ways, (c) be labelled with a specified health mark (unless the establishment is subject to a temporary hygiene derogation in which case the mark must not be used).

Sale of raw cows' milk as drinking milk

Raw cows' milk may only be sold as drinking milk in the following situations:

1) by the occupier of the premises where the cows are maintained to either (a) the ultimate consumer for consumption off-farm, (b) a temporary guest or visitor as part of a meal or refreshment, or (c) a distributor;
2) by a distributor if they are sold (a) in the same unopened containers as when received, (b) from a vehicle which is lawfully used as shop premises, and (c) direct to the ultimate consumer.

Conditions relating to all dairy establishments

Occupiers of dairy establishments shall do the following:

1) Ensure compliance with all aspects of the Regulations.
2) Carry out checks to ensure:
   a) that critical points in the establishment relative to the processes used there are identified;
   b) that methods for monitoring and controlling such critical points are established;
   c) that tests are carried out to detect antibiotics, pesticides, detergents or residues of substances having pharmacological or hormonal action, and other substances which are harmful to human health, which might alter the organoleptic characteristics of dairy products, or which might be harmful if consumption of residues exceeds permitted tolerance levels;
   d) that specified checks are carried out to detect the presence of added water;
   e) compliance with specified standards for animal health; and
   f) that samples taken for checking cleaning and disinfection methods are analysed and examined in approved laboratories.
3) Keep records of the checks specified in (2) for either 2 months from the 'best before' or 'use by' date of the product (in the case of dairy products stored chilled) or 2 years from the date of the record (in other cases) and allow for their inspection.
4) Notify the authorities if there is a serious health risk, and in the event of an immediate human health risk, withdraw from the market any similar products presenting the same risk.
5) Ensure that workers are given instruction and training in hygiene appropriate to any task undertaken by the worker.
6) If a purchaser of raw milk for re-sale as milk intended for processing, must meet specified requirements.

Microbiological requirements

Note: In the following tables, the letters *n, c, m* and *M* have the following meanings:

4.6

n = number of sample units comprising the sample;

c = number of sample units where the bacterial count may be between 'm' and 'M', the sample being considered acceptable if the bacterial count of the other sample is 'm' or less;

m = threshold value for the number of bacteria, the result being considered satisfactory if the number of bacteria in all samples does not exceed 'm';

M = maximum value for the number of bacteria, the result being considered unsatisfactory if the number of bacteria in one or more sample units is 'M' or more.

1. For milk being transferred from a production holding to a treatment or processing establishment:

	Plate count at 30°C (per ml)	Somatic cell count (per ml)	Staphylococcus aureus (per ml)			
			n	c	m	M
Raw milk:						
a) Intended for the production of heat-treated drinking milk, fermented milk, junket, jellied milk, flavoured milk or cream	max. 100 000	max. 400 000				
b) Intended for the manufacture of any dairy product not listed in (a)	max. 400 000 (from 1 January 1998, max. 100 000)	max. 500 000 (from 1 January 1998, max. 400 000)				
c) Intended for the production of any milk-based product (made with raw milk) which has not undergone any heat-treatment during its manufacture	max. 100 000	max. 400 000	5	2	500	2000
d) Raw goats', ewes' or buffaloes' milk intended for the production of heat-treated drinking milk or for the production of heat-treated milk-based products	max. 3 000 000 (from 1 December 1999, max. 1 500 000)					
e) Raw goats', ewes' or buffaloes' milk intended for the production of any milk-based product (made with raw milk) which has not undergone any heat-treatment during its manufacture	max. 1 000 000 (from 1 December 1999, max. 500 000)		5	2	500	2000
f) Raw cows' milk intended for the production of thermised milk if not subject to treatment within 36 hours of acceptance into processing establishment	max. 300 000					

2) For milk used for the manufacture of milk-based products and for milk-based products:

4.6

Product	Type of micro-organism or other requirement	Microbiological requirements				
			n	c	m	M
Thermised milk: (when used for the production of pasteurised, UHT or sterilised milk)	Plate count at 30°C	max. 100 000 per ml				
Heat-treated milk:						
1) Pasteurised (1)	a) Pathogenic micro-organisms	Absence in 25 g	5	0	–	–
	b) Coliforms (per ml)		5	1	0	5
	c) Plate count at 21°C (per ml)		5	1	5×10^4	5×10^5
	d) Plate count at 30°C (before any second heat-treatment)	max. 100 000 per ml				
2) Sterilised milk, UHT (2)	Plate count at 30°C	max. 100 per ml				
Milk-based products:						
1) Sterilised or UHT milk-based products which are in liquid or gel form (for storage at room temperature)	Plate count at 30°C (after incubation for 15 days at 30°C)	max. 100 per ml				
2) Cheese, other than hard cheese (3, 4)	*Listeria monocytogenes*	Absence in 25 g (5)	5	0		
3) Milk-based products other than cheese covered by (2)	*Listeria monocytogenes*	Absence in 1 g				
4) Milk powder (3, 4)	*Salmonella* spp.	Absence in 25 g	10	0		
5) Milk-based products other than (4) (3, 4)	*Salmonella* spp.	Absence in 25 g	5	0		
6) Cheese made from raw milk or from thermised milk (6)	a) *Staphylococcus aureus*		5	2	1000	10 000
	b) *Escherichia coli*		5	2	10 000	100 000
7) Soft cheese made from heat-treated milk	a) *Staphylococcus aureus*		5	2	100	1000
	b) *Escherichia coli*		5	2	100	1000
8) Fresh cheese, powdered milk, frozen milk-based products including ice-cream (6)	*Staphylococcus aureus*		5	2	10	100

Notes:
(1) Assessed by random sampling carried out in the treatment establishment
(2) Measured after 15 days in a closed container at 30°C (or, if necessary, 7 days at 55°C)
(3) If the standards for these products are exceeded, the products shall be excluded from human consumption and withdrawn from the market
(4) Testing of products is not compulsory for sterilised milk, preserved milk-based products and milk-based products where the heat treatment was applied after wrapping or packaging
(5) Sample to consist of five specimens of 5 g taken from different parts of the same product

4.6

(6) If the standards for these products are exceeded, there shall be a review of the implementation of the methods for monitoring and checking critical points in the processing establishment. If the value specified for M is exceeded in the case of cheese made from raw milk or from thermised milk or in the case of soft cheese, testing shall be carried out for the following and, if found, then all the batches of cheese involved shall be withdrawn from the market: presence of strains of enterotoxigenic *Staphylococcus aureus* or *Escherichia coli*, and, if necessary, the possible presence of staphylococcal toxins.

Specified heat treatments

	Temperature	Time	Required test	Result
Heat treated drinking milk:				
Thermised milk	57–68°C	15s	Phosphatase	Positive
Pasteurised milk (1)	min. 71.7°C	15 s	Phosphatase	Negative
UHT	min. 135°C	1 s	Peroxidase	Positive (2)
Milk-based products:				
Pasteurised cream				
a)	min. 63°C	30 min	Phosphatase	Negative
b)	min. 72°C	15 s	Phosphatase	Negative
Sterilised cream	min. 108°C	45 min		
UHT cream	min. 140°C	2 s		
Pasteurised ice-cream				
a)	min. 65.6°C	30 min		
b)	min. 71.1°C	10 min		
c)	min. 79.4°C	15 s		
Sterilised ice-cream	min. 148.9°C	2 s		

Notes:
(1) An equilvalent process may be used
(2) For 'high-temperature pasteurised milk', the peroxidase test will be negative

Storage and transport temperatures

	Maximum temperature (°C)
Raw milk	
Collected daily from a production holding, if not collected within 2 hours	8 (1)
Not collected daily from a production holding	6 (1)
During transport to a treatment or processing establishment (unless collected within 2 hours of milking)	10 (1)
At treatment establishment, unless heat-treated within 4 hours	6
Pasteurised milk	
After pasteurisation and during storage at treatment establishment	6
During transport in tanks, small containers or churns (except when intended for doorstep delivery)	6
During transport for delivery to a retail business (except when intended for doorstep delivery)	8
Any dairy product (except pasteurised milk)	
Not intended to be stored at ambient temperature	(2)

See next page for Notes

Notes:

(1) Approving authorities may permit higher temperatures when required for technological reasons relating to the manufacturer of certain milk-based products

(2) Temperature to be established by the manufacturer of the product as suitable to ensure its durability

4.6

Special labelling provisions

1) Dairy products must carry a health mark in an easily visible place (see Regulations for details).

2) Heat-treated milk and liquid milk-based products (unless packed in re-usable glass bottle) shall display:

   a) the nature of heat-treatment applied;

   b) an indication (or code) enabling date of last heat-treatment to be established; and

   c) for pasteurised milk, the storage temperature.

3) The following additional requirements are specified and should be clearly shown:

   a) for raw drinking milk intended for direct human consumption – 'raw milk';

   b) for milk-based products not having a thermisation or heat-treatment process – 'made with raw milk';

   c) for milk-based products where there is a heat-treatment – the nature of the treatment; or

   d) for milk-based products in which the growth of micro-organisms can occur – the 'use-by' or minimum durability date.

Ice-cream – heat treatment

Regulation: Ice-cream (Heat Treatment, etc.) Regulations 1959 (1959/734)
Amendments: Ice-cream (Heat Treatment, etc.)(Amendment) Regulations 1963 (1963/1083)
  Dairy Products (Hygiene) Regulations 1995 (1995/1086)
  Food Safety (General Food Hygiene) Regulations 1995 (1995/1763)

Note: These Regulations to not apply to ice-cream which is a milk-based product and subject to the requirements of the Dairy Products (Hygiene) Regulations 1995.

Definitions

Ice-cream: includes any similar commodity.

Ingredient: includes sugar and any whole egg, yolk or albumen (whether dried, frozen or preserved or not) but does not include colouring or flavouring materials or fruit, nuts, chocolate and other similar substances.

Mixture: a product which is capable of manufacture into ice-cream by freezing only.

4.6

Complete cold mix: a product which is capable of manufacture into a mixture with the addition of water only, is sent out by the manufacturer in airtight containers, and has been made by evaporating a liquid mixture which has already been submitted to heat treatment not less effective than that prescribed in these regulations, and to which, after treatment, no substance other than sugar has been added.

Requirements

1) Where a complete cold mix is used which is reconstituted with potable water and to which nothing is added other than sugar, colouring or flavouring materials, fruit, nuts, chocolate and other similar substances, the reconstituted mixture shall be converted into ice-cream within one hour of reconstitution.

2) In any other case (except for a mixture with a pH value of 4.5 or less used to make water ice or other similar frozen confection), the following apply:

   a) the mixture must not be kept above 45°F for more than 1 hour before heat treatment;

   b) one of the following heat treatments shall be used:
   Pasteurisation – 150°F for at least 30 minutes
   160°F for at least 10 minutes
   175°F for at least 15 seconds
   Sterilisation – 300°F for at least 15 seconds;

   c) after pasteurisation or sterilisation, the mix shall be reduced to a maximum of 45°F within 1.5 hours and shall be kept at such a temperature until freezing is begun unless:
   – it has been sterilised and immediately placed in sterile airtight containers under sterile conditions and remains unopened, and that, when opened, it is rapidly reduced to a maximum of 45°F until freezing is begun,
   – in the case of a mixture being added to a mixture (with a pH of 4.5 or less) for the preparation of a water ice or similar confection if the combined mixture is frozen within 1 hour of combination.

3) Ice cream shall be kept at a maximum temperature of 28°F prior to sale. If the temperature rises above 28°F it must be heat-treated again.

Meat

Regulation: Fresh Meat (Hygiene and Inspection) Regulations 1995 (1995/539)
Amendments: Food Safety (General Food Hygiene) Regulations 1995 (1995/1763)
Wild Game Meat (Hygiene and Inspection) Regulations 1995 (1995/2148)
Food Safety (Temperature Control) Regulations 1995 (1995/2200)
Colours in Food Regulations 1995 (1995/3124)
Fresh Meat (Hygiene and Inspection) (Amendment) Regulations 1995 (1995/3189)
Fresh Meat (Hygiene and Inspection) (Amendment) Regulations 1996 (1996/1148)

Note: These Regulations provide detailed controls on all aspects of the slaughter of specified animals and the cutting and sale of their meat. The provisions include requirements for licensing of slaughterhouses, cutting premises, cold stores, farmed game handling facilities and farmed game processing facilities; the supervision and control of premises (including the appointment by the Minister of Official Veterinary Surgeons (OVSs) and inspectors); conditions for the marketing of fresh meat; admission to and detention in slaughterhouses and farmed game processing facilities of animals and carcases. Only those points relating to conditions for the marketing of fresh meat are summarised below.

4.6

Definitions

Animals: the following food sources: (a) domestic animals of the following species – bovine animals (including buffalo of the species *Bubalus bubalis* and *Bison bison*), swine, sheep, goats and solipeds; and (b) farmed game.

Farmed game: wild land mammals which are reared and slaughtered in captivity, excluding (a) mammals of the family *Leporidae*, and (b) wild land mammals living within an enclosed territory under conditions of freedom similar to those enjoyed by wild game.

Meat: all parts of animals which are suitable for human consumption.

Fresh: as applied to meat, means all meat, including chilled or frozen meat, which has not undergone any preserving process and includes meat vacuum wrapped or wrapped in a controlled atmosphere.

Mechanically recovered meat: finely comminuted meat obtained by mechanical means from flesh-bearing bones (apart from (a) the bones of the head, and (b) the extremities of the limbs below the carpal and tarsal joints and , in the case of swine, the coccygeal vertebrae) and intended for approved meat products establishments, and includes mechanically separated meat.

Requirements

1) No person shall sell any fresh meat unless:
   a) it has been obtained from licensed premises (meeting specified conditions);
   b) it has been subject to specified ante-mortem health inspection and has been passed as fit for human consumption;
   c) it has been prepared under specified hygienic conditions;
   d) it comes from the body of an animal which has been subject to specified post-mortem health inspection which shows no evidence of disease or other abnormal condition (except for traumatic lesions incurred shortly before slaughter or localised problems which do not render the remainder of the carcase unfit for human consumption);
   e) it has been given a specified health mark;

199

4.6

f) it is accompanied during transport by specified documents;

g) if it has been stored in a cold store, the conditions met specified conditions;

h) if wrapped or packaged, the conditions met specified conditions;

i) if it is frozen, it was frozen under specified conditions;

j) if transported, the conditions met specified conditions;

k) if it is mechanically recovered meat, it has been handled in accordance with Directive 77/99.

2) No person shall sell any fresh meat which:

a) has been treated with natural or artificial colouring matter (except as required by the Regulations for health marking);

b) has been treated with ionizing or ultra-violet radiation;

c) is from male swine used for breeding or cryorchid or hermaphrodite swine unless the meat has been treated (as specified in Directive 77/99) and has a specified mark;

d) is from uncastrated male swine of a carcase weight exceeding 80 kg unless (i) an inspector has declared it to be free of sexual odours, or (ii) the meat has been treated (as specified in Directive 77/99) and has a specified mark;

e) is from animals to which a tenderiser has been administered.

3) Storage temperatures in cold stores:

fresh meat (after chilling) – max 7°C for carcases and cuts, max. 3°C for offal;

fresh meat (after freezing) – max. –12°C;

wild game meat (stored chilled) – max. 4°C for small wild game, max. 7°C for large wild game;

wild game meat (stored frozen) – max. –12°C.

Meat products

Regulation: Meat Products (Hygiene) Regulations 1994 (1994/3082)
Amendments: Food Safety (General Food Hygiene) Regulations 1995 (1995/1763)
  Food Safety (Temperature Control) Regulations 1995 (1995/2200)
  Food Labelling Regulations 1996 (1996/1499)

Note: These Regulations provide detailed controls on the handling, storing and marketing of meat products. The provisions include requirements for the approval of meat products premises, conditions for the handling, storing and marketing of meat products, certain conditions relating to all establishments, controls relating to prepared food obtained from raw materials of animal origin and for certain other products of animal origin. Only those points relating to conditions for the handling and marketing of meat products and certain conditions relating to all establishments are summarised below.

Definitions

4.6

Meat products: products for human consumption prepared from or with meat which has undergone treatment such that the cut surface shows that the product no longer has the characteristics of fresh meat, but not: (a) meat which has undergone only cold treatment, (b) minced meat, (c) mechanically recovered meat, and (d) meat preparations.

Meat preparations: preparations made wholly or partly from meat or minced meat (but excludes meat which has undergone only cold treatment) which: (a) have undergone a treatment other than one specified in these Regulations for meat products, (b) have been prepared by the addition of foodstuffs, seasonings or additives, or (c) have undergone a combination of the above.

Meat products premises: any industrial or non-industrial premises handling or storing meat products.

Handling: manufacturing, preparing, processing, packaging, wrapping, and re-wrapping.

Requirements

1) No person shall sell for human consumption from approved meat products premises any meat product manufactured in Great Britain unless:
   a) it has been handled and stored and the conditions met specified conditions;
   b) it has been prepared from raw materials which complied with specified requirements;
   c) it has been checked as specified.

2) No person shall sell for human consumption from approved meat products premises any meat product manufactured in Great Britain which is intended for consignment to an EEA State (other than Iceland) unless:
   a) it has been wrapped, packaged and labelled at manufacturing premises in accordance with specified requirements; and
   b) in the case of packaged meat products which cannot be stored at an ambient temperature, the packaging carries both a clear and legible indication of the temperature at which the product should be stored and transported, and the appropriate durability indication.

3) The occupier of an establishment shall take all necessary measures to ensure that the Regulations are complied with and shall carry out his own checks to ensure:
   a) that critical points in the establishment relative to the process used there are identified and acceptable to the enforcement authority;
   b) that methods for monitoring and controlling such critical points are established and acceptable to the enforcement authority;

4.6

c) that samples taken for checking cleaning and disinfection methods are analysed and examined in approved laboratories;
d) that records of the checks specified in (a)–(c) are kept for either 2 years or, in the case of meat products which cannot be stored at ambient, 6 months from the minimum durability date of the product;
e) that health marking is controlled and carried out properly.
4) Occupiers of establishments shall :
a) notify the authorities if there is a health risk, and in the event of an immediate health risk, withdraw from the market any similar products presenting the same risk;
b) ensure that workers are given instruction and training in hygiene appropriate to any task undertaken by them.

Minced meat and meat preparations

Regulation: Minced Meat and Meat Preparations (Hygiene) Regulations 1995 (1995/3205)

Notes:
a) These Regulations provide detailed controls on the handling, storing and marketing of minced meat and meat preparations. The provisions include requirements for the approval of premises, conditions for the marketing of minced meat and meat preparations, and duties of occupiers. Only those points relating to conditions for the marketing of minced meat and meat preparations, and duties of occupiers are summarised below.
b) The Regulations do not apply to:
i) premises producing or storing for direct sale to the final consumer from the premises;
ii) production of mechanically recovered meat; or
iii) production or sale of minced meat intended to be uses as a raw material for the production of sausage meat destined for inclusion in a meat product.

Definitions

Minced meat: meat which has been minced into fragments or passed through a spiral screw mincer and includes such meat to which not more than 1% salt has been added.

Meat preparation: meat to which foodstuffs, seasonings or additives have been added or which has undergone a treatment insufficient to modify its internal cellular structure and so alter its characteristics.

Meat product: any product, intended for human consumption, prepared from or with meat which has undergone treatment such that the cut surface shows that the product no longer has the characteristics of fresh meat, but does not include: (a) meat which has undergone only cold treatment, (b)

minced meat, (c) meat preparations, and (d) mechanically recovered meat.
Meat: parts of animals, excluding solipeds, or birds which are suitable
for human consumption and have been produced in a licensed
establishment (under fresh meat, poultry meat or wild game meat
regulations), or imported and examined under the Products of Animal
Origin (Import and Export) Regulations.

4.6

Requirements for minced meat
1) No person shall consign, or sell for consignment, to a relevant EEA
   State minced meat unless it is derived from meat of bovine animals,
   pigs, sheep or goats, and:
   a) it has been produced in approved premises;
   b) it has been prepared from striated muscle (other than heart
      muscle) including the adjoining fatty tissue, which, in the case of
      pig meat, has been examined for trichinae (or appropriately
      treated);
   c) it has been produced under specified conditions;
   d) it has been checked as specified;
   e) it has been labelled and given a health mark as specified;
   f) it has been wrapped or packaged and stored in a cold store as
      specified;
   g) it has been transported under specified conditions;
   h) it is accompanied during transport by specified documents;
   i) if it is derived from frozen or deep-frozen meat, it was deboned
      before freezing (except sheep meat and pig meat may be boned
      immediately before mincing) and had not been stored for more
      than (a) 18 months if beef or veal, (b) 12 months if sheep meat or
      goat meat, or (c) 6 months if pig meat;
   j) if it is derived from chilled meat, it has been minced within 6 days
      of slaughter or, if boned vacuum-packed beef or veal, 15 days;
   k) it has been cold treated within 1 hour of portioning and wrapping
      (except when production processes lower the internal
      temperature);
   l) if it is packaged and presented chilled, it is cooled to below
      2°C, or may incorporate limited frozen meat to accelerate
      refrigeration to less than 1 hour (but must be indicated on
      label);
   m) if it is packaged and presented deep-frozen, it meets specified
      conditions in the Quick Frozen Foodstuffs Regulations 1990;
   n) it has not been treated by ionising or ultra-violet radiation;
   o) if the label uses certain specified terms, it meets the related criteria
      (see 'Compositional criteria for meat', below).
2) No person shall sell for human consumption in Great Britain minced
   meat produced in the UK unless:
   a) it has been produced in approved premises or a registered

4.6

premises (licensed under meat product regulations or registered under the food premises registration regulations) and meets specified conditions;

b) it has been produced under certain specified conditions;
c) it has been subject to certain specified checks;
d) it has been wrapped or packaged as specified;
e) it has been stored under certain specified conditions;
f) if the label uses certain specified terms, it meets the related criteria (see 'Compositional criteria meat', below).

Requirements for meat preparations

1) No person shall consign, or sell for consignment, to a relevant EEA State a meat preparation unless:
   a) it has been produced in approved premises;
   b) if it is derived from pig meat, the meat has been examined for trichinae (or appropriately treated);
   c) if it is derived from frozen or deep-frozen meat the meat has been used, from time of slaughter, within (a) 18 months if beef or veal, (b) 12 months if sheep meat, goat meat, poultry meat, rabbit meat and farmed game meat, or (c) 6 months if any other meat;
   d) if it is packaged and presented chilled, it is cooled to below 2°C if it contains minced meat, 7°C if it contains fresh meat, 4°C if it contains poultry meat, and 3°C if it contains offal;
   e) if it is packaged and presented deep-frozen, it meets specified conditions in the Quick Frozen Foodstuffs Regulations 1990;
   f) it has been produced under specified conditions;
   g) it has been checked as specified;
   h) it has been labelled and given a health mark as specified;
   i) it has been wrapped or packaged and stored in a cold store as specified;
   j) it has been transported under specified conditions;
   k) it has not been treated by ionising radiation (except for medical purposes).

2) No person shall consign, or sell for consignment, to a relevant EEA State a meat preparation containing pre-prepared minced meat of bovine animals, pigs, sheep or goats unless the minced meat satisfies the requirements for minced meat given above.

3) No person shall sell for human consumption in Great Britain a meat preparation produced in the UK unless:
   a) it has been produced in approved premises or a registered premises (licensed under meat product regulations or registered under the food premises registration regulations) and meets specified conditions;
   b) it has been produced under certain specified conditions;

c) it has been subject to certain specified checks;
d) it has been stored under certain specified conditions in a cold store or, where unpackaged, in a licensed cold store;
e) it has not been treated by ionising radiation (except for medical purposes).

4.6

Additional requirements

Occupiers of any premises shall do the following:

1) Ensure compliance with all aspects of the Regulations.
2) Carry out checks to ensure that :
   a) critical points in the premises relative to production are identified and acceptable to the enforcement authority;
   b) methods for monitoring and controlling such critical points are established;
   c) samples taken for checking hygiene standards are analysed and examined in approved laboratories.
3) Keep records of the checks specified in (2) for either 2 months from the 'use by' date of the product (in the case of chilled minced meat or meat preparations) or 2 years (in other cases) and allow for their inspection.
4) Ensure that health marking is carried out properly.
5) Immediately notify the authorities if the analysis of a sample (or other information) reveals a health risk.
6) In the event of an imminent human health risk, withdraw from the market any similar products presenting the same risk.
7) Ensure that workers are given instruction and training in hygiene appropriate to any task undertaken by them.
8) Ensure that the packaging bears a storage temperature and either a 'use-by' date (chilled products) or 'best before' date (deep-frozen products).
9) Conduct and assess specified microbiological tests (see 'Microbiological criteria', below).

Compositional criteria for minced meat

Designations	Maximum fat content (%)	Maximum collagen content in meat protein (%)
'Lean minced', used in relation to meat of any permitted species	7	12
'Minced pure', used in relation to meat of bovine animals	20	15
'Minced', used in relation to meat of any permitted species and containing pig meat	30	18
'Minced', used in relation to sheep meat or goat meat	25	15

4.6 *Microbiological criteria*
    1) Minced meat:

	Lower threshold (per g)		Upper threshold (per g)		Microbic Limit
	Solid media	Liquid media	Solid media	Liquid media	
Aerobic mesophile bacteria	5×10^5	5×10^5	5×10^6	1.5×10^7	5×10^8
Escherichia coli	50	50	5×10^2	1.5×10^3	5×10^4
Staphylococcus aureus	10^2	10^2	5×10^3	3×10^3	5×10^4

Based on the above values, the minced meat is classified as follows:
    If all results at or below lower threshold: Category (a) – fully satisfactory;
    All results at or below upper threshold: Category (b) – acceptable if 2 in 5 or less tests on five consecutive samples produce results above lower threshold but at or below the upper threshold, otherwise unacceptable (except if this is only due to aerobic mesophile bacteria, a further sample from the same batch may be tested).
    Any result above the upper threshold: Category (c) – unacceptable.
    If any result above the microbic limit: minced meat to be regarded as presenting imminent risk to health.
    If a test for *Salmonella* shows a presence in 10 g, the quality shall be considered unacceptable.
    2) Meat preparations:

	Lower threshold (per g)		Upper threshold (per g)	
	Solid media	Liquid media	Solid media	Liquid media
Escherichia coli	5×10^2	5×10^2	5×10^3	1.5×10^4
Staphylococcus aureus	5×10^2	5×10^2	5×10^3	3×10^3

Based on the above values, the meat preparation is classified as follows:
    If all results at or below lower threshold: Category (a) – fully satisfactory.
    All results at or below upper threshold: Category (b) – acceptable if 2 in 5 or less tests (for *Escherichia coli*) or 1 in 5 or less tests (for *Staphylococcus aureus*) on five consecutive samples produce results above lower threshold but at or below the upper threshold, otherwise unacceptable.

Any result above the upper threshold: Category (c) – unacceptable.

If any result above the microbic limit: minced meat to be regarded as presenting imminent risk to health.

If a test for *Salmonella* shows a presence in 1 g, the quality shall be considered unacceptable.

4.6

Poultry meat, farmed game bird meat and rabbit meat

Regulation: Poultry Meat, Farmed Game Bird Meat and Rabbit Meat (Hygiene and Inspection) Regulations 1995 (1995/540)
Amendments: Food Safety (General Food Hygiene) Regulations 1995 (1995/1763)
Food Safety (Temperature Control) Regulations 1995 (1995/2200)

Note: These Regulations provide detailed controls on all aspects of the slaughter of poultry, farmed game meat and rabbits and the cutting and sale of their meat. The provisions include requirements for licensing of slaughterhouses, cutting premises, cold stores and re-wrapping centres; the supervision and control of premises; and prohibitions and conditions of slaughter. Only those points relating to conditions for the marketing of fresh meat are summarised below.

Definitions
Poultry: means domestic fowls, turkeys, guinea fowls, ducks and geese.
Farmed game birds: birds, including ratites, but excluding poultry, which are not generally considered domestic but which are bred, reared and slaughtered in captivity.
Rabbit: a domestic rabbit.

Requirements
1) Subject to a number of exceptions, no person shall sell any poultry meat, farmed game meat or rabbit meat unless:
   a) it has been obtained from licensed premises (meeting specified conditions);
   b) it has been subject to specified pre-slaughter health inspection, has been passed as fit for human consumption and has then been slaughtered;
   c) it has been chilled and prepared under specified hygienic conditions;
   d) it comes from a body of a bird or rabbit which has been subject to specified post-mortem health inspection which shows no evidence of disease or other abnormal condition (except for localised problems which have been removed and which do not render the remainder of the carcase unfit for human consumption);
   e) it has been given a specified health mark;
   f) it is accompanied during transport by specified documents;

4.6

g) if it has been stored, the conditions met specified conditions;

h) if wrapped or packaged, the conditions met specified conditions;

i) if transported, the conditions met specified conditions.

2) Subject to a number of limited exceptions, no person shall sell any fresh meat from poultry, farmed game or rabbit which:

a) has been treated with an antibiotic or tenderiser;

b) has been marked with a colourant (except as required by the Regulations);

c) has been treated with a preservative other than a permitted preservative;

d) has been chilled immediately after evisceration and post-mortem health inspection by an immersion technique which does not meet the specified conditions;

e) in the case of poultry meat, has not been eviscerated;

f) in the case of rabbit meat or farmed game meat, has been treated with ionising or ultra-violet radiation.

3) Fresh meat, after chilling, shall be kept at or below 4°C, or, after freezing, at or below –12°C.

Wild game meat

Regulation: Wild Game Meat (Hygiene and Inspection) Regulations 1995 (1995/2148)

Note: These Regulations provide detailed controls on all aspects of the processing of wild game meat intended for consignment, or sale for consignment, to another EEA State (excluding Iceland). The provisions include requirements for licensing of wild game processing facilities; the supervision and control of premises (including the appointment by the Minister of Official Veterinary Surgeons (OVSs), inspectors and Plant Inspection Assistants (PIAs)); and conditions for the marketing of wild game meat. Only those points relating to conditions for the marketing of wild game meat are summarised below.

Definitions

Wild game: both (a) wild land mammals which are hunted (including wild mammals living within an enclosed area under conditions of freedom similar to those enjoyed by wild game), and (b) wild birds.

Large wild game: wild ungulates.

Small wild game: wild mammals of the *Leporidae* family, and wild birds intended for human consumption.

Wild game meat: all parts of wild game which are suitable for human consumption and which have not undergone any preserving process other than chilling, freezing, vacuum wrapping or wrapping in a controlled atmosphere.

Requirements

1) Subject to a number of exceptions, no person shall consign, or sell for consignment, to a relevant EEA State wild game meat unless:
   a) it has been obtained from supervised licensed premises or from a cold store or re-wrapping centre (meeting specified conditions);
   b) it comes from wild game which (i) has been killed by hunting, (ii) does not come from a region subject to restrictions specified in certain EC Directives, (iii) immediately after killing has been prepared (in a specified way) and transported to licensed premises under satisfactory hygiene conditions within a reasonable time and at a temperature which allows for post-mortem inspection, and (iv) has been handled under specified hygienic conditions;
   c) it comes from a body of a wild game animal which has been subject to specified post-mortem health inspection and approved fit for human consumption;
   d) in the case of small wild game, a representative sample of such killed game from the same source or batch has undergone a specified inspection;
   e) it has been given a specified health mark;
   f) it is accompanied during transport by specified documents;
   g) if it has been stored, the conditions met specified conditions;
   h) if wrapped or packaged, the conditions met specified conditions;
   i) if transported, the conditions met specified conditions.
2) Subject to a number of limited exceptions, no person shall consign, or sell for consignment, to a relevant EEA State:
   a) wild game meat which has been condemned as unfit for human consumption or has been treated with an antibiotic or tenderiser;
   b) wild game meat which has been treated with ionizing or ultra-violet radiation.
   c) wild game meat which has been treated with colouring other than used for health marking;
   d) wild game meat obtained from animals which have ingested substances which are likely to make the meat dangerous or harmful to human health;
   e) offal or large wild game declared fit for human consumption unless it has undergone treatment specified in Directive 77/99 (as amended);
   f) unskinned or unplucked and uneviscerated small wild game which has been frozen or which has not been handled and stored separately from fresh meat and skinned or plucked wild game meat;
   g) unskinned large wild game, unless: (i) it has been killed by hunting and it does not come from a region subject to restrictions specified in certain EC Directives, (ii) it has been visually

4.6

inspected by an inspector, (iii) such viscera as originally accompanied the game to the premises have undergone post-mortem inspection, (iv) it has a veterinary health certificate (of a specified form), (v) it has been cooled to (and transported to a wild game processing facility) either between –1°C and +7°C (if transported within 7 days of the post-mortem inspection), or –1°C and +1°C (if transported within 15 days of the post-mortem inspection).

3) Temperatures to be met after post-mortem inspection and for transport:

carcases and cuts of large wild game meat (chilled): max. 7°C;
carcases and cuts of small wild game (chilled): max. 4°C;
wild game meat (stored frozen): max. –12°C.

Egg products

Regulation: Egg Products Regulations 1993 (1993/1520)
Amendments: Food Safety (General Food Hygiene) Regulations 1995 (1995/1763)
Food Labelling Regulations 1996 (1996/1499)

Note: These Regulations include requirements for the preparation of egg products; approval of establishments used for the manufacture of egg products; the supervision of production in approved establishments; storage, transport and packaging of egg products, and the marking of egg products. Only those points relating to requirements for the preparation and storage of egg products are summarised below.

Definitions
Egg: an egg laid by a hen, duck, goose, turkey, guinea fowl or quail.
Egg products: products obtained from eggs, their various components or mixtures thereof, after removal of the shell and outer membranes, intended for human consumption, and includes such products when partially supplemented by other foodstuffs and additives and such products when liquid, concentrated, crystallised, frozen, quick-frozen or dried but does not include finished foodstuffs.
Whole egg: a mixture of yolk and albumen.
Cracked eggs: eggs with a damaged but unbroken shell, with intact membranes.

General
1) Egg products which are a mixture of egg products obtained from more than one species cannot be sold, except egg products (using certain specified eggs) may be used at the establishment where they were obtained if they comply with certain hygiene requirements (see Regulations).

210

2) Egg products must comply with specified hygiene and processing requirements (see below).

3) Egg products prepared in Great Britain must have been prepared in an approved establishment, approval being given by the local authority. Egg products from other EC countries must have been prepared in establishments which meet the equivalent EC standards.

Hygiene, processing and storage

1) Only non-incubated eggs which are fit for human consumption may be used, and their shells must be fully developed and contain no breaks. Cracked eggs may be used if delivered direct from the packing centre or farm to the heat treatment establishment and broken there as quickly as possible.

2) The egg products shall be removed from the shell by a technique in which the eggs are broken avoiding contamination of the egg contents and allows inspection of the contents of individual eggs. They shall not be obtained by centrifuging or crushing.

3) Pasteurisation conditions are specified as follows:
   a) whole egg or yolk – min. 64.4°C for min. 2 minutes 30 seconds (or an equivalent process) followed by cooling as quickly as possible to below 4°C (except temperature may be kept above 4°C for the purpose of dissolving sugar or salt in whole egg or yolk). Pasteurised product should pass a specified α-amylase test.
   b) albumen – heat-treated by a process designed to take account of the likely microbiological contamination levels in the untreated albumen so as to meet microbiological criteria (see (4)).

4) For each batch of heat treated egg products the following microbiological criteria shall be met:
   a) salmonella – absence in 25 g or 25 ml of egg products;
   b) mesophilic aerobic bacteria – $M = 10^5$ in 1 g or 1 ml;
   c) enterobacteriaceae – $M = 10^2$ in 1 g or 1 ml;
   d) *Staphylococcus aureus* – absence in 1 g of egg products;
   where M = maximum value for the number of bacteria; the result is considered unsatisfactory if the number of bacteria in one or more sample units is M or more.

5) Samples of egg products shall meet the following criteria:
   a) 3-OH-butyric acid – max. 10 mg/kg in the dry matter of the unmodified egg product;
   b) lactic acid – max. 1000 mg/kg of egg product dry matter (applicable only to the untreated egg product);
   c) succinic acid – max. 25 mg/kg of egg product dry matter.

6) If heat treatment is not carried out immediately after breaking, egg product shall be stored frozen or at a temperature of not more than 4°C (for up to 48 hours maximum, except in the case of ingredients to be desugared).

4.6

7) The quantity of egg shell remains, egg membrane and any other particles in the egg product shall not exceed 100 mg/kg of egg product.

8) Maximum storage temperatures are specified:
   a) deep frozen egg products, −18°C;
   b) other frozen products, −12°C;
   c) chilled products, +4°C.

Fishery products

Regulations: Food Safety (Fishery Products) Regulations 1992 (1992/3163)
    Food Safety (Fishery Products on Fishing Vessels) Regulations 1992 (1992/3165)
Amendments: Food Safety (Fishery Products)(Import Conditions and Miscellaneous Amendments) Regulations 1994 (1994/2783)
    Food Safety (General Food Hygiene) Regulations 1995 (1995/1763)
    Food Safety (Fishery Products and Live Bivalve Molluscs and Other Shellfish) (Miscellaneous Amendments) Regulations 1996 (1996/1547)
    Food Labelling Regulations 1996 (1996/1499)

Note: These Regulations include requirements for the placing of fishery products, aquaculture products and processed bivalve molluscs and other shellfish on the market; approval of factory vessels and establishments; registration of fishing vessels on which shrimps or molluscs are processed by cooking; registration of wholesale and auction markets that are not establishments; and sales by fishermen of small quantities of fishery products. Only those points relating to requirements for the placing of fishery products, aquaculture products and processed bivalve molluscs and other shellfish on the market are summarised below.

Definitions

Fishery products (for the Fishery Products Regulations): (a) all seawater or freshwater animals, including their roes; and (b) parts of such animals, except in circumstances where they (i) are combined (in whatever way) with other foodstuffs, and (ii) comprise less that 10% of the total weight of the combined foodstuffs, but excluding aquatic mammals, frogs and aquatic animals covered by Community Acts other than Directive 91/493, and parts of such mammals, frogs and aquatic animals.

Fishery products (for the Fishery Products on Fishing Vessels Regulations): all seawater or freshwater animals or parts thereof, including their roes, but excluding aquatic mammals, frogs and aquatic animals covered by other Community Acts.

Aquaculture products: (a) all fishery products born or raised in controlled conditions until placed on the market as a foodstuff, and (b) all seawater fish, freshwater fish or crustaceans caught in their natural environment when juvenile and kept until they reach the desired commercial size for human consumption, other than fish or crustaceans of commercial size caught in their natural environment and kept alive to be sold at a later date, if they are merely kept alive without any attempt being made to increase their size or weight.

Placing on the market: in relation to fishery products for human consumption, the holding for sale, exposing for sale, displaying for sale, selling, consigning, delivering or any other associated activity of marketing but not either a sale to a final consumer or a sale by a fisherman of a small quantity within a local market in specified circumstances (see Regulations for details).

Requirements for fishery products
1) Subject to a number of exceptions, no person shall place on the market for human consumption any fishery products, unless:
   a) if they have been handled on board a British Islands fishing vessel, the hygiene requirements of the Food Safety (Fishery Products on Fishing Vessels) Regulations 1992 (1992/3165) were met (See Regulations for details);
   b) if either they have been handled on board a British Islands factory vessel, or if they were landed in the British Islands, during and after landing, the specified Community hygiene requirements as given in the Food Safety (Fishery Products) (Derogations) Regulations 1992 (1992/1507) were met (see Regulations for details);
   c) at establishments on land in the British Islands, they have been handled and, where appropriate, packaged, prepared, processed, frozen, defrosted, stored hygienically and inspected in accordance with the requirements as given in the Food Safety (Fishery Products) (Derogations) Regulations 1992 (1992/1507) (see Regulations for details);
   d) they have been subject to specified health control and monitoring (including organoleptic checks, parasite checks, chemical checks for TVB-N, TMA-N and histamine, and microbiological analyses (see also Commission Decision 93/51 on the microbiological criteria applicable to the production of cooked crustaceans and molluscan shellfish));
   e) they have been appropriately packaged (meeting specified requirements);
   f) they are a consignment (or part of one) which bears a specified identification mark (unless subject to an exemption specified in the Regulations);
   g) they have been stored and transported under satisfactory conditions of hygiene (meeting specified requirements);
   h) if imported, prior to importation they satisfied the appropriate requirements of the Food Safety (Fishery Products) (Import Conditions and Miscellaneous Amendments) Regulations 1994

4.6

(1994/2783) (see Regulations for details);

   i) they meet the additional requirements specified below.

2) A fishery product should be gutted as soon as possible (from a technical and commercial viewpoint) after the product has been caught or landed.

3) No person shall place on the market for human consumption any fishery products which are aquaculture products, unless:
   a) they have been slaughtered under appropriate specified conditions of hygiene;
   b) they have not been soiled with earth, slime or faeces;
   c) if they were not processed immediately after being slaughtered, they have been kept chilled;
   d) they meet the requirements of (1) (c)–(i) above.

4) No person shall place on the market for human consumption any fishery products which are processed bivalve molluscs or other shellfish, unless:
   a) they met, prior to processing, the appropriate requirements of the Food Safety (Live Bivalve Molluscs and Other Shellfish) Regulations 1992 (1992/1508) (see Regulations for details);
   b) they meet the requirements of (1) (c)–(i) above.

5) A person with control over the survival conditions of a fishery product which is to be placed on the market alive shall ensure that it is at all times kept under conditions most suitable for its survival.

6) No person shall place on the market any of the following:
   a) poisonous fish of the following species: *Tetraodontidae*, *Molidae*, *Diodontidae* or *Canthigasteridae*;
   b) fishery products containing biotoxins including ciguatera or muscle-paralysing toxins.

Requirements for fishery products on fishing vessels

No person shall carry out on a fishing vessel any commercial operation in relation to fishery products unless certain specified hygienic conditions are fulfilled. (See the Food Safety (Fishery Products on Fishing Vessels) Regulations for details.)

Requirements for imported fishery products

The Food Safety (Fishery Products) (Import Conditions and Miscellaneous Amendments) Regulations 1994, as amended, impose restrictions on imports from Community and third countries. See Regulations for full details but the main requirement is that no person shall import any fishery product for human consumption unless either, they have satisfied the relevant Community hygiene requirements, or, if they originate in a third country, they meet specified requirements. Additional conditions are imposed on imports from certain third countries.

Live bivalve molluscs and other shellfish

4.6

Regulations: Food Safety (Live Bivalve Molluscs and Other Shellfish) Regulations
1992 (1992/3164)
Food Safety (Live Bivalve Molluscs and Other Shellfish) (Import Conditions and
Miscellaneous Amendments) Regulations 1994 (1994/2782)
Amendments: Food Safety (General Food Hygiene) Regulations 1995 (1995/1763)
Food Safety (Fishery Products and Live Bivalve Molluscs and Other Shellfish)
(Miscellaneous Amendments) Regulations 1996 (1996/1547)
Food Labelling Regulations 1996 (1996/1499)

Note: These Regulations include requirements for designation of certain
seawaters and brackish waters from which live bivalve molluscs may be
taken; designation of areas unsuitable for production or harvesting;
approval of dispatch centres and purification centres (and restrictions on
their operation) placing live bivalve molluscs and other shellfish on the
market; and monitoring by Ministers and food authorities. Only certain of
these points are summarised below.

Definitions

Bivalve mollusc: filter-feeding lamellibranch molluscs.

Other shellfish: live echinoderms, tunicates or marine gastropods.

Production area: any sea, estuarine or lagoon area containing either
natural deposits of bivalve molluscs, or sites used for the cultivation of
bivalve mollusc, including relaying areas, from which live bivalve
molluscs are taken.

Relaying area: any area of sea, estuary or lagoon within boundaries
clearly marked and indicated by buoys, posts or any other fixed means
and which is used exclusively for the natural purification of live bivalve
molluscs.

Purification centre: an establishment with tanks fed by naturally clean
sea water or sea water that has been cleaned by appropriate treatment,
in which live bivalve molluscs are placed for the time necessary to
remove microbiological contamination, so making them fit for human
consumption.

Dispatch centre: any on-shore or off-shore installation for the reception,
conditioning, washing, cleaning, grading or wrapping of either live
bivalve molluscs or other shellfish, or both, for human consumption.

Requirements

1) Subject to certain exemptions, no person shall place on the market for
   immediate human consumption any live bivalve molluscs or other
   shellfish, unless:
   a) they originate for a designated Class A, B or C production area
      (unless they are pectinidae which are not aquaculture products, or
      imported live bivalve molluscs) (see (3) below);
   b) they have been harvested, kept and transported to any approved

215

4.6

dispatch centre, approved purification centre or approved relaying area to which they are thereafter transferred in accordance with specified requirements;

c) they have (where necessary) been relaid in accordance with specified requirements;

d) they have (where appropriate) been purified or subject to intensive purification at an approved purification centre;

e) they meet the standards specified in (2) below;

f) any wrapping meets specified requirements;

g) they have been stored and transported under satisfactory conditions in accordance with specified requirements;

h) they are a consignment (or part of one) which bears a specified health mark (unless subject to an exemption specified in the Regulations);

i) they have been handled hygienically;

j) if they are imported, prior to importation they have met the requirements of the Food Safety (Live Bivalve Molluscs and Other Shellfish) (Import Conditions and Miscellaneous Amendments) Regulations 1994.

2) Live bivalve molluscs and other shellfish for immediate human consumption shall meet the following:

a) possession of visual characteristics associated with freshness and viability, including shells free of dirt, an adequate response to percussion, and normal amounts of intravalvular liquid;

b) either less than 300 faecal coliforms or less than 230 E.coli per 100 g of mollusc flesh and intravalvular liquid;

c) no salmonella in 25 g of mollusc flesh;

d) no toxic or objectionable compounds occurring naturally or added to the environment in such quantities that the calculated dietary intake exceeds the permissible daily intake (PDI) or that the taste of the molluscs is impaired;

e) maximum 80 μg per 100 g total paralytic shellfish poison (PSP) content in the edible parts of molluscs;

f) absence of diarrhoetic shellfish poison (DSP) in the edible parts of molluscs.

3) Production areas are classified as follows:

Class A: an area from which live bivalve molluscs (meeting requirements of (2) above) can be gathered for direct human consumption.

Class B: an area from which live bivalve molluscs may: (a) be gathered but only marketed after treatment in a purification centre or after relaying, (b) be used for relaying (and followed by purification), or (c) be heat-treated in an approved fishery products establishment. Ninety per cent of molluscs from these areas must not exceed 6000 faecal coliforms per 100 g of flesh, or 4600 E.coli per 100 g of flesh.

Class C: an area from which live bivalve molluscs may be gathered but marketed only after (a) a relaying period of at least 2 months (whether or not followed by purification), (b) approved intensive purification, or (c) given an approved heat treatment in an approved fishery products establishment. Prior to relaying, etc., molluscs must not exceed 60 000 faecal coliforms per 100 g flesh.

4.6

Requirements for imported live bivalve molluscs or other shellfish
The Food Safety (Live Bivalve Molluscs and Other Shellfish) (Import Conditions and Miscellaneous Amendments) Regulations 1994, as amended, impose restrictions on imports from Community and third countries. See Regulations for full details, but the main requirement is that no person shall import any live bivalve molluscs or other shellfish for human consumption unless either they have satisfied the relevant Community hygiene requirements, or, if they originate in a third country, they meet specified requirements. Additional conditions are imposed on imports from certain third countries.

Imports and exports of food

A General
Regulation: Imported Food Regulations 1984 (1984/1918)
Amendments: Channel Tunnel (Amendment of Agriculture, Fisheries and Food Import Legislation) Order 1990 (1990/2371)
    Products of Animal Origin (Import and Export) Regulations 1992 (1992/3298)

Note: For most products of animal origin, the Imported Food Regulations 1984 have largely been superseded by the Products of Animal Origin (Import and Export) Regulations 1992 (see next section). The latter Regulations disapply most aspects of the Imported Food Regulations for relevant products. The Ministry is in the process of consultations with regard to introducing new Imported Food Regulations which would significantly simplify the controls.

Requirements
The main provisions of the Regulations are as follows.
1) No person shall import any food intended for sale for human consumption:
   a) which has been rendered injurious to health,
   b) which has been examined and found not to be fit for human consumption,
   c) which is otherwise unfit for human consumption or is unsound or unwholesome, or
   d) has used any food covered by (a)–(c) in its preparation.
2) If on examination a food is found not to comply with the

4.6

requirement of (1), action may be taken as if the food was failing to comply with food safety requirements and appropriate action taken (as provided in Section 9 of the Food Safety Act 1990).

3) For bulk lard, fresh meat or a meat product (i.e. products prepared from fresh meat but excluding: vitamin concentrates containing meat, pharmaceutical products containing meat, gelatine, rennet, meat products of which meat is not a principal ingredient and which do not contain fragments of meat), additional provisions apply, including:

   a) procedures for obtaining examination by an independent veterinary expert when an authorised officer believes that the food does not comply,

   b) a requirement that the product bears a specified and recognised health mark,

   c) specified wrapping, packing and transporting requirements which include the following maximum temperatures:

      i)   rabbit meat/offals, hare meat/offals, poultry meat/offals, 4°C;
      ii) carcases and cuts not listed in (i), 7°C (chilled), –12°C (frozen);
      iii) offals not listed in (i), 3°C;
      iv) meat products, temperature as specified on the label.

   d) A requirement that any fresh meat which is derived from domestic bovine animals (including buffalo), swine, sheep, goats, solipeds or poultry or meat products prepared from such meat (but with certain exemptions) must be accompanied by a valid health certificate as specified in the Regulations.

B Animal and poultry products

Regulations: Importation of Animal Products and Poultry Products Order 1980 (1980/14)

Products of Animal Origin (Import and Export) Regulations 1992 (1992/3298)

Amendments: Importation of Animal Products and Poultry Products (Amendment) Order 1980 (1982/948)

Channel Tunnel (Amendment of Agriculture, Fisheries and Food Import Legislation) Order 1990 (1990/2371)

Food Safety (Fishery Products)(Import Conditions and Miscellaneous Amendments) Regulations 1994 (1994/2783)

Importation of Animal Products and Poultry Products (Amendment) Order 1994 (1994/2920)

Note: The 1980 Order given above provides a general requirement for licensing of imports and contain enforcement provisions. The 1992 Regulations provide for specific controls so as to implement EC requirements for checks on animal products (both imports and exports) and they disapply the enforcement provisions of the 1980 Order for those products covered by the 1992 Regulations.

Requirements – 1980 Order

4.6

The landing of any animal product or poultry product from a place outside Great Britain is prohibited except under the authority of a license in writing issued by the Minister and in accordance with the conditions of that license. The license may be general or specific and may contain conditions which are considered necessary to prevent the spreading of any animal or bird disease into or within Great Britain. Substances subject to the provisions of the Medicines Act 1968 and animal pathogens or carriers defined in the Importation of Animal Pathogens Order 1980 are exempt from the requirement.

Requirements – 1992 Regulations

1) It is an offence to export to another Member State any product of animal origin covered by the Directives listed in (4) below unless it complies with the requirements of those Directives (for products not covered by one of the listed Directives, any requirements of the Member State must be met).

2) For intra-Community trade, checking requirements are specified for both border inspection posts (for imported products which are not transported by regular, direct means of transport linking two geographical points of the Community) and to provide the power to inspect at the place of destination or during transport (when there is a suspicion of an irregularity).

3) For third country trade, the Regulations specify approved border inspection posts, import procedures, and action to be taken when consignments pose a risk to health or are illegal.

4) The Directives which must be complied with in (1) are:

    64/433   on health problems affecting intra-Community trade in fresh meat,

    71/118   on health problems affecting trade in fresh poultry meat,

    72/461   on health problems affecting intra-Community trade in fresh meat,

    77/99   on health problems affecting intra-Community trade in meat products,

    80/215   on animal health problems affecting intra-Community trade in meat products,

    85/397   on health and animal health problems affecting intra-Community trade in heat-treated milk,

    88/657   laying down the requirements for the production of, and trade in, minced meat, meat in pieces of less than 100 g and meat preparations,

    89/437   on hygiene and health problems affecting the production and the placing on the market of egg products,

    91/67   concerning the animal health conditions governing the placing on the market of aquaculture animals and products,

4.6

91/492   laying down the health conditions for the production and the placing on the market of live bivalve molluscs,

91/493   laying down the health conditions for the production and the placing on the market of fishery products,

91/494   on animal health conditions governing intra-Community trade in and imports from third countries of fresh poultry meat,

91/495   concerning public health and animal health problems affecting the production and placing on the market of rabbit meat and farmed game meat.

Perishable foods (international)

Regulation: The International Carriage of Perishable Foodstuffs Regulations 1985 (1985/1071)

Amendment (relevant to food): International Carriage of Perishable Foodstuffs (Amendment) Regulations 1991 (1991/425)

For amendments relating to transport, see 1989/1185, 1990/2033, 1991/969, 1991/2571, 1992/2682 and 1993/1589

Note: The Regulations govern the international carriage of perishable foodstuffs by rail or road transport (and includes cases when the carriage involves sea crossings of less than 150 km). The following, taken from Annexes 2 and 3 of the ATP (see page 21), are the maximum temperatures during carriage (and, in the case of (1) and (2) below, on loading and unloading).

1) Quick (deep)-frozen and frozen foodstuffs (1):	
Ice-cream	−20°C
Frozen or quick (deep)-frozen fish, fish products, molluscs and crustaceans and all other quick (deep)-frozen foodstuffs	−18°C
All frozen foodstuffs (except butter)	−12°C
Butter	−10°C
2) Deep-frozen and frozen foodstuffs to be immediately further processed at destination:	
Butter	(2)
Concentrated fruit juice	(2)
3) Chilled foods:	
Red offal	3°C (3)
Butter	6°C
Game	4°C
Milk (raw or pasteurised) in tanks, for immediate consumption	4°C (3)
Industrial milk	6°C (3)
Dairy products (yoghurt, kefir, cream and fresh cheese) (4)	4°C (3)
Fish, molluscs and crustaceans (5)	(6)
Meat products (7)	6°C
Meat (other than red offal)	7°C
Poultry and rabbits	4°C

See next page for Notes

Notes:
(1) A brief rise in the temperature of the surface of the foods of max. 3°C in a part of the load is permitted during certain operations (e.g. defrosting of evaporator)
(2) May be permitted to gradually rise in temperature to temperature specified in transport contract but must not be higher than that specified for the product under (3) in the table
(3) In principle, duration of carriage should not exceed 48 hours
(4) Fresh cheese means a non-ripened (non-matured) cheese which is ready for consumption shortly after manufacture and which has a limited conservation period
(5) Other than smoked, salted, dried or live fish, live molluscs and live crustaceans
(6) Must always be carried in melting ice
(7) Except for products stabilised by salting, smoking, drying or sterilisation

4.6

4.7 Weights and measures

Prescribed weights and quantities

Orders: Weights and Measures Act 1963 (Intoxicating Liquor) Order 1988 (1988/2039)
Weights and Measures Act 1963 (Miscellaneous Foods) Order 1988 (1988/2040)
Amendments: Weights and Measures Act 1963 (Various Foods) (Amendment) Order 1990 (1990/1550)
Weights and Measures (Metrication)(Miscellaneous Goods) (Amendment) Order 1994 (1994/2868)

Given below are details of the prescribed weights and quantities for prepacked food and drink (for beer, cider, wine and intoxicating liquor, see the Intoxicating Liquor Order for details). Certain exemptions may be allowed for certain packages (e.g. several small packets in an outer wrapping). The Orders should be consulted for full details. The Orders also provide details of foods which (a) should be prepacked only if marked with an indication of quantity either by net weight or by capacity measurement, (b) shall be pre-packed only if the container is marked with an indication of quantity by number, and (c) exemptions to these requirements. Check Orders for details.

Barley kernels, pearl barley, rice (including ground rice and rice flakes), sago, semolina, and tapioca: except when 75 g or less, more than 10 kg
125 g , 250 g, 375 g , 500 g or a multiple of 500 g.
Biscuits and shortbread: includes wafers (except when not cream-filled), rusks, crispbreads, extruded flatbread, oatcakes and matzos (except when produced and packed on premises where sold), except when 85 g or less, more than 5 kg
100 g, 125 g , 150 g, 200 g, 250 g, 300 g or a multiple of 100 g.

4.7

Bread: whole loaves of bread in any form (except breadcrumbs) including fancy loaves, milk loaves, rolls and baps and sliced prepacked bread, except when 300 g or less
400 g or a multiple of 400 g.

Cereal breakfast foods: in flake form, (but not cereal biscuit breakfast foods), except when 50 g or less, more than 10 kg
125 g , 250 g, 375 g , 500 g, 750 g, 1 kg, 1.5 kg or a multiple of 1 kg.

Cocoa and chocolate products:
a) all cocoa products (including drinking chocolate) except when less than 50 g, more than 1 kg
50 g, 75 g , 125 g , 250 g, 500 g, 750 g, 1 kg;
b) the following chocolate in bar or tablet form – plain, gianduja nut, milk, gianduja nut milk, white, filled, cream, skimmed milk – except when less than 85 g , more than 500 g
85 g , 100 g, 125 g , 150 g, 200 g, 250 g, 300 g, 400 g, 500 g.

Coffee, coffee mixtures and coffee bags: (for coffee bags, the weight relates to the contents), except when less than 25 g , more than 5 kg
57 g, 75 g , 113 g, 125 g , 227 g, 250 g, 340 g, 454 g, 500 g, 680 g, 750 g or a multiple of 454 g or of 500 g.

Coffee extracts and chicory extracts: consisting of solid and paste coffee and chicory products, except when 25 g or less, more than 10 kg
50 g, 100 g, 200 g, 250 g (for mixtures of coffee extracts and chicory extracts only), 300 g (for coffee extracts only), 500 g, 750 g, 1 kg, 1.5 kg, 2 kg, 2.5 kg, 3 kg or a multiple of 1 kg.

Dried fruits: apples (and apple rings), apricots, currants, dates, figs, muscatels, nectarines, peaches, pears (and pear rings), prunes, raisins, sultanas and dried fruit salad, except when 75 g or less, more than 10 kg
125 g , 250 g, 375 g , 500 g, 1 kg, 1.5 kg, 7.5 kg or a multiple of 1 kg.

Dried vegetables: beans, lentils, peas (includes split peas), except when 100 g or less, more than 10 kg
125 g , 250 g, 375 g , 500 g, 1 kg, 1.5 kg, 7.5 kg or a multiple of 1 kg.

Edible fats:
a) butter, margarine, butter/margarine mixtures and low fat spreads, except when 25 g or less, more than 10 kg, and
b) dripping and shredded suet, lard and compound cooking fat (and substitutes thereof), and solidified edible oil (except in gel form), except when less than 5 g , more than 10 kg
50 g, 125 g , 250 g, 500 g or a multiple of 500 g up to 4 kg, or a multiple of 1 kg from 4 kg to 10 kg.

Flour and flour products: flour of bean, maize, pea, rice, rye, soya bean, wheat, and cake flour (except cake or sponge mixtures), cornflour (except blancmange or custard powders) and self-raising flour, except when 50 g or less, more than 10 kg
125 g , 250 g, 500 g or a multiple of 500 g (except for cornflour add 375 g , 750 g).

Honey: all honey (except chunk honey and comb honey), except when
   less than 50 g

57g, 113 g, 227 g, 340 g, 454 g, 680 g or a multiple of 454 g.

Jam, marmalade and jelly preserves: (except diabetic jam and
   marmalade), except when less than 50 g

57 g, 113 g, 227 g, 340 g, 454 g, 680 g or a multiple of 454 g.

Milk:

   a) other than in a returnable container, except when 50 ml or less

189 ml, 200 ml, 250 ml, 284 ml, 500 ml, 750 ml or a multiple of 284 ml or
of 500 ml;

   b) in a returnable container, except when 50 ml or less

1/3pt, 1/2pt or a multiple of 1/2pt,

or 200 ml, 250 ml, 500 ml, 750 ml or a multiple of 500 ml.

Molasses, syrup and treacle: except when less than 50 g

57 g, 113 g, 227 g, 340 g, 454 g, 680 g or a multiple of 454 g.

Oat products: namely flour of oats, oatflakes and oatmeal, except when
   50 g or less, more than 10 kg

125 g , 250 g, 375 g , 500 g, 750 g, 1 kg, 1.5 kg or a multiple of 1 kg.

Pasta: except when 50 g or less

125 g , 250 g, 375 g , 500 g or a multiple of 500 g.

Potatoes:

   a) for prepacked potatoes, except when less than 5 g , more than 25 kg

500 g, 750 g, 1 kg, 1.5 kg, 2 kg, 2.5 kg or a multiple of 2.5 kg up 15 kg, 20
kg or 25 kg;

   b) for potatoes 175 g or over each, they may be sold by number (with a
      statement of the minimum weight of each potato).

Salt: except when 100 g or less

125 g , 250 g, 500 g, 750 g, 1 kg, 1.5 kg or a multiple of 1 kg up to 10 kg,
12.5 kg, 25 kg, 50 kg.

Sugar: except when 100 g or less, more than 5 kg

125 g , 250 g, 500 g, 750 g, 1 kg, 1.5 kg, 2 kg, 2.5 kg, 3 kg, 4 kg, 5 kg.

Tea: includes tea sold in tea bags but excludes instant tea, except when
   25 g or less, more than 5 kg

50 g, 125 g , 250 g, 500 g, 750 g, 1 kg, 1.5 kg, 2 kg, 2.5 kg, 3 kg, 4 kg, 5 kg
(except for tea, other than tea in a tea bag, packed in tins or glass or
wooden containers, add 100 g, 200 g, 300 g).

Average weights

Regulation: Weights and Measures (Packaged Goods) Regulations 1986 (1986/2049)
Amendments: Weights and Measures (Quantity Marking and Abbreviations of
Units) Regulations 1987(1987/1538)
Weights and Measures (Packaged Goods)(Amendment) Regulations 1992
(1992/1580)
Weights and Measures (Packaged Goods and Quantity Marking and Abbreviation
of Units) (Amendment) Regulations 1994 (1994/1852)

4.7 These Regulations provide the full details of the procedures to be followed to satisfy the requirements contained in Part V of the Weights and Measures Act 1985. It is impossible to summarise all aspects of the detailed Regulations, and packers and importers are advised to consult the Regulations or the Code of Practical Guidance for Packers and Importers published by HMSO and issued under the Act.

Products covered by the average system
The Regulations specify that for most food products packed to a predetermined constant quantity with a declared weight of between 5 g and 10 kg (5 ml to 10 l) the average weight system will be used and that the products should conform to the Regulations. Outside these specified limits products should be packed to a minimum weight system.

The following products have prescribed limits different from those given above except when marked with the 'e' mark and packed to satisfy the requirements of 'e' marked goods:

> Biscuits (includes wafers, rusks, crispbreads, extruded flatbread, oatcakes and matzos) and shortbread (except as specified below): 50 g to 10 kg.
>
> Bread (as defined under Prescribed Weights) in single loaf form: 300 g to 10 kg (per loaf).
>
> Cheese of the following types: processed cheese, cheese spread, cottage cheese and the following natural cheese types – Caerphilly, Cheddar, Cheshire, Derby, Double Gloucester, Dunlop, Edam, Gouda, Lancashire, Leicestershire and Wensleydale: 25 g to 10 kg.
>
> Cocoa products and chocolate products: 50 g to 10 kg.
>
> Edible fats as follows: butter, margarine, butter/margarine mixtures and low fat spreads, dripping and shredded suet, lard and compound cooking fat and substitutes, solidified edible oil (except in gel form): 5 g to 20 kg.
>
> Herbs and spices: 25 g to 4 kg.
>
> Potato crisps and other similar snack foods: 25 g to 10 kg.
>
> Poultry (or any part of poultry) of any description or food consisting substantially of poultry: 5 g to 20 kg.
>
> Sugar: 50 g to 10 kg.
>
> Sugar confectionery and chocolate confectionery (except as specified below): 50 g to 10 kg.
>
> Alcoholic beverages (except certain wines, sparkling wines, spirits and liqueurs listed in the Weights and Measures (Intoxicating Liquor) Order 1988 to which the general limits apply): 5 ml to 5 l.
>
> Edible oil in liquid or gel form: 5 ml to 20 l.
>
> Single portion vending machine beverage packs (to include any beverage mentioned below): 25 g to 10 kg or 25 ml to 10 l.

Single portion catering packs (to include any foods mentioned below): 25 g to 10 kg or 25 ml to 10 l.

4.7

The following products are excluded from the average weight system:
Bath chaps, meat pies, meat puddings and sausage rolls.
Single cooked sausages in natural casings weighing less than 500 g.
Sausage meat products other than in sausage form weighing less than 500 g.
Poultry pies, fish pies.
Cheese (except as specified above).
Flour confectionery (except when consisting of or including uncooked pastry or shortbread) and including bun loaves, fruit loaves, malt loaves and fruited malt loaves.
Fresh fruit and vegetables.
Ice lollies, water ices and freeze drinks, ice-cream.
Single toffee apples.
Soft drinks in a syphon.
Sugar confectionery and chocolate confectionery of the following types: Easter eggs, figurines of sugar or of chocolate, rock or barley sugar in sticks or novelty shapes, or a collection of these articles
Shortbread in a piece or pieces each weighing 200 g or more.
Biscuits in packs weighing less than 100 g sold on the premises on which they are packed.

Three rules for packers
The primary duty of a packer or importer is to ensure that an Inspector's reference test is passed. He can do this by ensuring that his packages comply with three rules:
1) The actual contents of the packages shall be not less, on average, than the nominal quantity (i.e. the weight or volume declared).
2) Not more than 2.5% of the packages may be below the nominal quantity (a negative error) by an amount larger than the tolerable negative error (TNE) specified for the nominal quantity. Packages below this weight are known as 'non-standard'.
3) No package may have a negative error larger than twice the TNE specified. Packages below this weight are known as 'inadequate'. It is generally accepted in the Code of Guidance that in practice this rule will be satisfied if no more than one package in 10 000 is likely to be inadequate (i.e. 0.01%) .

The following table gives the specified TNEs for different nominal quantities:

4.7

Nominal quantity (g or ml)	Tolerable negative error	
	% of nominal quantity (1)	g or ml
5–50	9	–
50–100	–	4.5
100–200	4.5	–
200–300	–	9
300–500	3 ·	–
500–1000	–	1.5
1000–10 000	1.5	–
10 000–15 000	–	150
Above 15 000	1	–

Note: (1) To be rounded up to nearest 1/10 g or ml

Details and records of checks that are conducted to satisfy the three rules for packers must be kept for 12 months.

Use of the 'e' mark
The 'e' mark (shown in the figure below) may only be used when the nominal quantity is in the range 5 g to 10 kg (or 5 ml to 10 l). The mark shall be at least 3 mm high, placed in the same field of vision as the weight/volume statement, and indelible, clearly legible and visible under normal conditions of purchase. The mark is not obligatory but when used is a guarantee, recognised throughout the EEC, that the goods to which it is applied have been packed in accordance with the relevant EEC Directive.

However, when used, packers must notify the relevant authorities of their use of the 'e' mark before the end of the day on which the packages are marked. An inspector may give a notice to a packer relieving him of this requirement where the 'e' mark is used regularly or frequently. The following cases should be considered:

1) Product packed in the UK for internal use only: no benefit is gained by using the 'e' mark unless the product falls outside the range of weight or volume specified in the Regulations for that product, yet is still within the 5 g – 10 kg (or 5 ml to 10 l) range of the EEC Directive. In this special case it would allow the packages to be packed to the average system.

2) Product packed in the UK for export to the EEC: guarantees that products meet the EEC Directive.

3) Imported goods:
    a) from EEC – use of the mark relieves the importer from checking the goods on import, or from obtaining certain specified documentation;
    b) from outside the EEC – use of the mark is still possible but the importer must notify the authorities of the type, nominal quantity and place of import of the goods. Certain checks may then be required.

4.7

Units of measurement

Regulation: Units of Measurement Regulations 1986 (1986/1082)
Amendment: Units of Measurement Regulations 1994 (1994/2867)

The Regulations generally require the International System of Units (SI) to be used for measurement. This includes the kilogram (kg) and metre (m). The unit of volume is derived from the metre and is 1 cubic decimetre. The following exceptions relevant to food are provided:
    1) between 1 October 1995 and 1 January 2000, the following may be used:
        a) fluid ounce and pint – for beer, cider, water, lemonade and fruit juice in returnable containers;
        b) ounce and pound – for goods for sale loose from bulk.
    2) from 1 October 1995, the following may be used:
        pint – for dispense of draught beer or cider and for milk in returnable containers.

Weight and capacity marking

Regulation: Weights and Measures (Quantity Marking and Abbreviations of Units) Regulations 1987 (1987/1538)
Amendments: Weights and Measures (Quantity Marking and Abbreviations of Units) (Amendment) Regulations 1988 (1988/627)
Weights and Measures (Packaged Goods and Quantity Marking and Abbreviation of Units) (Amendment) Regulations 1994 (1994/1852)

These Regulations govern the methods of marking allowed for weights, capacity and other measures.
The following are included in the requirements:
    1) Weight/capacity marking shall be in metric (except as provided in (2)) but may in addition include a supplementary indication in imperial units. The following units and abbreviations are permitted:
        kilogram (kg), gram (g);
        litre (l or L), centilitre (cl or cL), millilitre (ml or mL).
    2) In addition to metric units, returnable containers used
        a) for milk may be marked by reference to the pint (pt or pts); and

4.7

    b) for beer, cider, water, lemonade and fruit juice may be marked by reference to the pint (pt or pts) and the fluid ounce (fl oz or fl ozs).

3) Weight/capacity may be in words or figures (but if expressed in words, the unit shall also be expressed in words).

4) If weight is gross weight, then 'gross' or 'including container' to be stated.

5) The words 'net' or 'gross' must not be abbreviated.

6) A metric quantity shall not be expressed as a vulgar fraction.

7) Size of marking: the minimum size of figures to be used is determined by the mass or capacity of the contents as follows (catchweight products are exempt from this requirement)

    not exceeding 50 g/5 cl: 2 mm

    over 50 g/5 cl but not over 200 g/20 cl: 3 mm

    over 200 g/20 cl but not over 1 kg/1 l: 4 mm

    over 1 kg/1 l: 6 mm.

Appendix 1

A Regulations for food standards, hygiene, additives and contaminants

Notes:
1) This list includes the year and number of the Regulations or Rules which apply in Scotland and Northern Ireland. The requirements are usually the same although the administrative arrangements may be different. Where an asterisk (*) is printed, any Regulations or Rules which do exist, are not directly comparable, and the Regulations or Rules given in Parts 3 and 4 should be consulted.
2) This list does not include Regulations which affect only the penalties or enforcement of the listed Regulations. In particular, the following Regulations are not listed:

Milk and Dairies (Revision of Penalties) Regulations 1982 (1982/1703)
Foods (Revision of Penalties) Regulations 1982 (1982/1727)
Food (Revision of Penalties) Regulations 1985 (1985/67)
Milk and Dairies (Revision of Penalties) Regulations 1985 (1985/68)
Food (Revision of Penalties and Mode of Trial) (Scotland) Regulations 1985 (1985/1068)
Food Safety Act 1990 (Consequential Modifications) (England and Wales) Order 1990 (1990/2486)
Food Safety Act 1990 (Consequential Modifications) (No. 2) (Great Britain) Order 1990 (1990/2487)
Food Safety Act 1990 (Consequential Modifications)(Scotland) Order 1990 (1990/2625)
Food Safety (Exports) Regulations 1991 (1991/1476)
Food (Forces Exemptions) (Revocations) Regulations 1992 (1992/2596)

3) Certain administrative Regulations are not included in this book and the more important are as follows:

A1

Authorised Officers (Meat Inspection) Regulations 1987 (1987/133)

Food Safety (Enforcement Authority) (England and Wales) Order 1990 (1990/2462)

Food Safety (Sampling and Qualifications) Regulations 1990 (1990/2463)

Detention of Food (Prescribed Forms) Regulations 1990 (1990/2614)

Food Safety (Improvement and Prohibition – Prescribed Forms) Regulations 1991 (1991/100)

Food Premises (Registration) Regulations 1991 (1991/2825) (as amended by 1993/2022)

Part 1   Listing of Regulations before the Food Safety Act 1990

Subject	England and Wales		Scotland		Northern Ireland	
	Regulations	Amendments	Regulations	Amendments	Rules	Amendments
Milk and dairies (general)	1959/277	1992/3143	*		*	
Ice-cream (heat treatment)	1959/734	1963/1083	*		*	
		1995/1086				
		1995/1763				
Arsenic	1959/831	1960/2261	1959/928	1960/2344	1961/98	1964/172
		1973/1052		1963/1461		1972/275
		1973/1340		1972/1489		1973/197
		1975/1486		1973/1039		1973/466
		1992/1971		1973/1310		1984/406
		1995/3202		1984/1518		1992/416
				1992/1971		
				1995/3202		
Mineral hydrocarbons	1966/1073	1995/3187	1966/1263	1992/2597	1966/200	1992/463
				1995/3187		1996/50
Sugar products	1976/509	1980/1834	1976/946	1980/1889	1976/165	1980/28
		1980/1849		1981/137		1981/193
		1982/255		1982/410		1981/305
		1995/3124		1995/3124		1982/311
		1995/3187		1995/3187		1996/50
		1996/1499		1996/1499		
Cocoa and chocolate	1976/541	1980/1833	1976/914	1980/1888	1976/183	1981/193
		1980/1834		1980/1889		1981/194
		1980/1849		1981/137		1981/305
		1982/17		1982/108		1982/349
		1984/1305		1984/1519		1984/407
		1995/3187		1995/3187		1996/50
		1995/3267		1995/3267		1996/53
		1996/1499		1996/1499		1996/338
Honey	1976/1832	1980/1849	1976/1818	1981/137	1976/387	1981/305
		1996/1499		1996/1499		1996/383
Drinking milk	1976/1883	1992/3143	*		1977/8	1992/559
		1995/1086				1995/201
Erucic acid	1977/691	1982/264	1977/1028	1982/18	1977/135	1982/184
Fruit juices and nectars	1977/927	1979/1254	1977/1026	1979/1641	1977/182	1980/28
		1980/1849		1981/137		1981/305
		1982/1311		1982/1619		1983/48

230

A1

Subject	England and Wales		Scotland		Northern Ireland	
	Regulations	Amendments	Regulations	Amendments	Rules	Amendments
Condensed and dried milk	1977/928	1991/1284 1995/236 1995/3267 1995/3187 1996/1499 1980/1849 1982/1066 1986/2299 1989/1959 1995/3187 1996/1499	1976/1027	1992/1284 1995/236 1995/3267 1995/3187 1996/1499 1981/137 1982/1209 1987/26 1989/1975	1977/196	1991/251 1995/106 1996/53 1996/383 1981/305 1983/26 1987/65 1989/430
Coffee and coffee products	1978/1420	1980/1849 1982/254 1987/1986 1995/3187 1996/1499	1979/383	1981/137 1982/409 1987/2014 1995/3187 1996/1499	1979/51	1980/28 1981/305 1982/298 1988/23 1996/50
Lead	1979/1254	1985/912 1992/1971 1995/3124 1995/3267	1979/1641	1985/1438 1922/1971 1995/3124 1995/3267	1979/407	1981/194 1985/163 1992/416 1996/53
Animal and poultry products (import)	1980/14	1982/948 1990/2371 1994/2920	1980/14	1982/948 1990/2371 1994/2920	*	
Chloroform	1980/36		1980/289		1980/75	
Jam and similar products	1981/1063	1982/1700 1983/1211 1988/2112 1989/533 1990/2085 1995/3123 1995/3124 1995/3187 1996/1499	1981/1320	1982/1779 1983/1497 1988/2084 1989/581 1990/2180 1995/3123 1995/3124 1995/3187 1996/1499	1982/105	1982/398 1983/265 1988/433 1989/152 1990/205 1990/388 1996/50 1996/383
Poultry meat (water content)	1984/1145		1983/1372	1984/1576	1982/386	
Meat products and spreadable fish products	1984/1566	1986/987 1995/3123 1995/3124 1995/3187 1996/1499	1984/1714	1986/1288 1995/3123 1995/3124 1995/3187 1996/1499	1984/408	1996/50 1996/383
Imported food	1984/1918	1990/2371 1992/3298	1985/913		1984/1917 (1)	1991/475
Natural mineral waters	1985/71	1996/1499	1985/71	1996/1499	1985/120	1996/383
Perishable foods (international carriage)	1985/1071	1991/425	1985/1071	1991/425	1985/1017 (1)	1991/425
Caseins and caseinates	1985/2026	1989/2321 1996/1499	1986/836	1990/1 1996/1499	1986/40	1990/37 1996/383
Materials and articles in contact with food	1987/1523	1994/979	1987/1523	1994/979	1987/432	1994/174
Olive oil	1987/1783	1992/2590	1987/1783	1992/2590	1987/431	1993/9

A1

Subject	England and Wales		Scotland		Northern Ireland	
	Regulations	Amendments	Regulations	Amendments	Rules	Amendments
Tetrachloroethylene in olive oil	1989/910		1989/837		1989/261	
Milk (protection of designations)	1990/607	1995/3267 1996/1499	1990/816	1995/3267 1996/1499	1990/103	1996/53 1996/383
Sardines (preserved)	1990/1084		1990/1139		1990/194	
Tryptophan in food	1990/1728		1990/1792		1990/329	

Note: (1) Statutory Instrument

Part 2 Listing of Regulations after the Food Safety Act 1990

Subject	England, Wales and Scotland		Northern Ireland	
	Regulations	Amendments	Rules	Amendments
Irradiation of food	1990/2490		1992/172	
Quick-frozen foodstuffs	1990/2615	1994/298 1996/1499	1990/455	1994/52
Meat (residue limits)	1991/2843	1993/990 1994/2465	1992/39	1995/97
Tin	1992/496		1992/166	
Fishery products (derogations)	1992/1507	1995/1763	1992/296	1993/51 1995/113 1995/360
Live bivalve molluscs (derogations)	1992/1508	1995/1763	1992/295	1993/52 1995/360
Flavourings	1992/1971	1994/1486 1996/1499	1992/416	1994/270 1996/383
Food additives (labelling)	1992/1978	1995/3123 1995/3124 1995/3187 1996/1499	1992/417	1996/50 1996/383
Organic products	1992/2111	1993/405 1994/2286	1992/2111 (1)	1993/405 (1) 1994/2286 (1)
Plastic materials and articles in contact with food	1992/3145	1995/360 1996/694	1993/173	1995/107 1996/164
Fishery products	1992/3163	1994/2783 1995/1763 1996/1499 1996/1547	1993/51	1995/113 1995/360 1996/264
Live bivalve molluscs and other shellfish	1992/3164	1994/2782 1995/1763	1993/52	1995/112 1995/360
Fishery products (on fishing vessels)	1992/3165	1994/2783 1995/1763 1996/1547	1993/53	1995/113 1995/360 1996/264
Aflatoxin	1992/3236	1996/1499	1993/31	
Products of animal origin (import and export)	1992/3298		1993/304	1995/113
Egg products	1993/1520	1995/1763 1996/1499	1993/329	1995/360
Extraction solvents	1993/1658	1995/1440	1993/330	1995/263

Subject	England, Wales and Scotland		Northern Ireland	
	Regulations	Amendments	Rules	Amendments
Drinking water in containers	1994/743		1994/184	
Pesticide residues (crops, food and feedingstuffs)	1994/1985	1995/1483	1995/33	1995/461 1996/527
Tuna and bonito (preserved)	1994/2127		1994/425	1996/383
Live bivalve molluscs (import conditions)	1994/2782	1996/1547 1996/1499	1995/112	1996/264 1996/383
Fishery products (import conditions)	1994/2783	1995/1763 1995/2200 1996/1547	1995/113	1995/360 1995/377 1996/264
Meat products (hygiene)	1994/3082	1995/539 1995/1763 1995/2200 1995/3205 1995/1499		
Infant formula and follow-on formula	1995/77	1995/3267 196/1499	1995/85	1996/53 1996/383
Fresh meat (hygiene and inspection)	1995/539	1995/1763 1995/2148 1995/2200 1995/3124 1995/3189 1996/1148		
Poultry meat, farmed game bird meat, rabbit meat (hygiene and inspection)	1995/540	1995/1763 1995/2200 1995/2148	1955/396	
Dairy products (hygiene)	1995/1086	1996/1499 1996/1499 1996/1699	1995/201	1995/360 1996/53 1996/287 1996/383
Food hygiene (general)	1995/1763	1995/2200	1995/360	1995/377
Wild game meat (hygiene and inspection)	1995/2148	1995/3205		1996/286
Temperature control	1995/2200	1995/3205 1996/1499	1995/377	
Spreadable fats	1995/3116		1996/47	
Sweeteners	1995/3123		1996/48	
Colours	1995/3124		1996/49	
Miscellaneous food additives	1995/3187		1996/50	
Bread and flour	1995/3202	1996/1499 1996/1501	1996/51	1996/385
Minced meat and meat preparations	1995/3205			
Specified bovine material	1996/1192		1996/361	
Food labelling	1996/1499		1996/383	
Food labelling (lot marking)	1996/1502		1996/384	

Note: (1) Statutory Instrument; (2) These Regulations do not apply to Scotland – see next table (Part 3)

A1

Part 3   Regulations specific to Scotland

These Regulations are specific to Scotland and are not directly comparable to Regulations in force in England and Wales. Full details of these Regulations are not given in this guide.

Subject	Regulation	Amendments
Ice-cream (heat treatment)	1948/960	1948/2271
		1960/2108
		1963/1101
		1995/1372
		1995/1763
Food hygiene (additional controls)	1959/413	1959/1153
		1961/622
		1966/967
		1978/173
		1995/1763
Drinking milk	1976/1888	1992/3136
		1995/1372
Milk (labelling)	1983/938	1990/2508
		1992/3136
		1995/1372
Milk and dairies	1990/2507	1992/3136
		1995/1372
Dairy products (hygiene)	1995/1372	1996/1499

Part 4   Regulations and rules specific to Northern Ireland

These Regulations (usually published as Statutory Rules) are specific to Northern Ireland and are not directly comparable to Regulations in force in England and Wales. Full details of these Rules are not given in this guide.

Subject	Rules	Amendments
Ice-cream (heat treatment)	68/13	95/201
		93/360
		96/53
Import of poultry products	1965/175	1968/106
Landing of animal products	1970/145	1972/113
		1976/324
Imported animals (diseases)	1981/115 (1)	
Meat inspection	1984/402	1991/5
Pesticide residues (national levels)	1995/32	1995/460
		1996/526
Mechanically recovered meat	1995/470	
Beef (emergency controls)	1996/132	1996/404
Fresh meat (beef controls)	1996/404	1996/506

Note: (1) Statutory Instrument

B Regulations for trade descriptions and weights

A1

Subject	England, Wales and Scotland		Northern Ireland	
	Regulations	Amendments	Rules	Amendments
Trade descriptions (Exemption 1)	1972/1886		1972/1886 (1)	
Trade descriptions (Exemption 10)	1976/260		1976/260 (1)	
Prescribed weights (intoxicating liquor)	1988/2039	1990/1550	1989/164	1994/320
Prescribed weights (miscellaneous foods)	1988/2040	1990/1550 1994/2868	1989/69	1990/395 1995/230
Average weight controls	1986/2049	1987/1538 1992/1580 1994/1852	1990/410	1991/320 1992/485 1994/321 1995/229
Weight marking and abbreviations	1987/1538	1988/627 1994/1852	1991/320	1995/229

Note: (1) Statutory Instrument

Appendix 2

A Reports of the Food Standards Committee (FSC)

1 Foods and additives

Antioxidants	1953, 1954, 1963
Beer	1977
Bread and flour	1960,1974
Bread, milk	1959
Cheese	1982
Cheese (processed cheese and cheese spread)	1949, 1956
Cheese (processed cheese)	1949
Cheese (hard, soft and cream cheeses)	1962
Coffee mixtures	1951
Colouring matters	1954, 1955, 1964
Cream	1950, 1967, 1982
Cream (artificial and synthetic)	1951
Curry powder	1948
Emulsifying and stabilising agents	1956
Fish cakes	1951
Fish paste	1950
Fish and meat pastes	1965
Flavouring agents	1965
Gelatine	1950
Ice-cream	1957
Infant formulae (artificial feeds for young infants)	1981
Jams and marmalade	1952
Jams and other preserves	1968
Margarine and other table spreads	1981
Margarine, vitaminisation of	1954
Meat pies	1963
Meat, canned	1962
Meat products	1980
Meat products, offals in	1972
Milk, condensed	1969, 1973
Milk, dried	1962
Mince	1983
Mineral oil	1962
Novel protein foods	1974
Preservatives	1959, 1960

Saccharin tablets	1952
Salt, iodisation of	1950
Sausages	1956
Soft drinks	1959
Soft drinks – Part I: fruit juices and nectars	1975
Soft drinks – Part II: soft drinks	1976
Soups	1968
Suet	1952
Tomato ketchup	1948
Vinegars	1971
Water in food	1978
Yogurt	1975

2 Miscellaneous

Claims and misleading descriptions	1966, 1980
Date marking of food	1972
Food labelling	1964, 1980
Pre-1955 compositional orders	1970

3 Trace elements

Arsenic	1950, 1951, 1955
Copper	1951, 1956
Fluorine	1954, 1957
Lead	1951, 1954
Tin	1954
Zinc	1954

B Reports of the Food Additives and Contaminants Committee (FACC)

Aldrin and dieldrin	1967
Antioxidants	1965, 1966, 1971, 1974
Arsenic	1984
Asbestos	1979
Azodicarbonamide	1968
Beer, additives and processing aids used in the production of	1978
Bread and flour, additives to	1971
Bulking aids	1980
Cheese	1982
Colouring matter in food	1979
Cream	1982
Cyclamates	1966, 1967
Emulsifiers and stabilisers	1970, 1972
Enzyme preparations	1982
Flavourings in food	1976
Flavour modifiers	1978
Further classes of food additives	1968
Infant formulae	1981
Leaching of substances from packaging materials into food	1970

A2

Lead in food	1975
Liquid freezants	1972, 1974
Metals in canned foods	1983
Mineral hydrocarbons in food	1975
Modified starches	1980
Nitrites and nitrates in cured meat and cheese	1978
Packaging	1970
Preservatives	1972
Solvents	1966, 1974, 1978
Sorbic acid	1977
Sulphur dioxide in canned peas	1977
Sweeteners	1982

C Reports of the Food Advisory Committee (FdAC)

1.	Skimmed milk with non-milk fat	1984
2.	Additives to cloud soft drinks	1985
3.	Coated and ice-glazed fish products	1987
4.	Colouring matter	1987
5.	Old compositional orders (includes mustard, curry powder, tomato ketchup, suet, salad cream and ice-cream)	1989
6.	Response to comments on the review of colouring matter	1989
7.	Potassium bromate	1990
8.	Mineral hydrocarbons	1990
9.	Saccharin	1990
10.	Food labelling and advertising	1990
11.	Emulsifiers and stabilisers	1992
12.	Use of additives in food specially prepared for infants and young children	1992

Note: Since 1992, the Food Advisory Committee has not issued formal reports. Annual reports have however been published.

Appendix 3

Codes of practice

Codes of practice often provide additional guidance as to the interpretation of legislation. This Appendix lists those which have significant legal authority. For a more complete listing of about 400 codes, readers are advised to consult the *Listing of Codes of Practice Applicable to Foods* published by the Institute of Food Science and Technology, 1993.

A LAJAC

For full details of these codes, see the *Journal of the Association of Public Analysts* in the issues stated below. For the terms of reference of the LAJAC, see the *Journal* (1963, vol. 1(2), p. 50).

1) Use of the word 'chocolate' in flour confectionery (1963, vol. 1(2), p. 49): min. 3% dry non-fat cocoa solids in the moist crumb (does not apply if called 'chocolate flavoured').
2) Labelling of Brandy (1963, vol. 1(4), p. 101): brandy must come from grapes; 'Cognac' and 'Armagnac' must come from those regions.
3) Crab meat content in Norwegian canned crab products (1963, vol 1(4), p. 105).
4) Canned fruit and vegetables. (Now replaced by new Code of Practice; see LACOTS below.)
5) Canned beans in tomato sauce (1965, vol. 3(4), p. 111): covers beans and sauce; standards for total solids, tomato solids and sugar (according to can size).
6) Marzipan, almond paste and almond icing (1969, vol. 7(2), p. 40): min. 23.5% almond substance (no other nut ingredient) and min. 75% of remainder to be solid carbohydrate sweetening matter.
7) Bun loaves, chollas and buttercream (1971, vol. 9(4), p. 107):
Bun loaf – min. 5% liquid whole egg and/or fat; min. 10% egg and or fat and/or sugar
Chollas – normally same as for bun loaf

A3

Buttercream (not 'butter cream' or 'butter-cream') – min. 22.5% butter fat and no other added fat.

B LACOTS

The following codes have been prepared and adopted by the relevant industrial associations in consultation with LACOTS:

1) UK Association of Frozen Food Producers: Code of Practice for the Breadcrumb Covering of Scampi; 1980.
2) Dairy Trade Federation: Code of Practice for the Composition and Labelling of Yogurt; 1983.
3) British Meat Manufacturers Association: Code of Practice on the Labelling of Re-formed Cured Meat Products; 1985.
4) Honey Importers and Packers Association: Code of Practice for the Importation, Blending, Packaging and Marketing of Honey; 1986.
5) British Fruit and Vegetable Canners' Association: Code of Practice on Canned Fruit and Vegetables; 1986 (replaced LAJAC Code No. (4).
6) Fish and Meat Spreadable Products Association: Code of Practice for the Composition of Meat and Fish Pastes and Pâtés; 1986.
7) Grain and Feed Trade Association: Rice Standards Section Code of Practice for Rice; 1992.

C NMCU

The following codes have been prepared to help manufacturers meet the requirements of the average weight system. They were prepared in consultation with the NMCU and have been approved by the Secretary of State for Trade and Industry:

1) National Association of Master Bakers, Confectioners and Caterers and the Scottish Association of Master Bakers: Code of Practical Guidance of Small Bakers; 1982.
2) Federation of Bakers: Code of Practice on Bread Weight Checking for Plant Bakers; amended version 1985.
3) British Soft Drinks Council: The Average System of Quantity Controls on Packages: Code of Practice for the Soft Drinks Industry; 1983.
4) Brewers' Society: The Average System of Quantity Control: Code of Practice for Beer Packaging; 1983.
5) Cake and Biscuit Alliance: Code of Practical Guidance for the Biscuit Industry; 1983.
6) British Poultry Federation: Code of Practice for the Average Weighing of Whole Poultry and Poultry Portions; 1985.

D Enforcement codes

The following codes have been issued under Section 40 of the Food Safety Act 1990 and are available from HMSO:

1) Responsibility for Enforcement of the Food Safety Act 1990, 1991.
2) Legal Matters, 1991.
3) Inspection Procedures – General, 1991.
4) Inspection, Detention and Seizure of Suspect Food, 1991.
5) The Use of Improvement Notices, 1991 (revised 1994).
6) Prohibition Procedures, 1991.
7) Sampling for Analysis or Examination, 1991.
8) Food Standards Inspections, 1991 (revised 1996).
9) Food Hygiene Inspections, 1991 (revised 1995).
10) Enforcement of the Temperature Control Requirements of Food Hygiene Regulations, 1991 (revised 1994).
11) Enforcement of the Food Premises (Registration) Regulations, 1991.
12) Quick Frozen Foodstuffs – Division of Enforcement Responsibilities, 1991 (revised 1994).
13) Enforcement of the Food Safety Act 1990 in Relation to Crown Premises, 1992.
14) Enforcement of the Food Safety (Live Bivalve Molluscs and other Shellfish) Regulations 1992, 1994.
15) Enforcement of the Food Safety (Fishery Products) Regulations 1992, 1994.
16) Enforcement of the Food Safety Act 1990 in Relation to the Food Hazard Warning System, 1993.
17) Enforcement of the Meat Products (Hygiene) Regulations 1994, 1994.
18) Enforcement of the Dairy Products (Hygiene) Regulations 1995, 1995.
19) Qualifications and Experience of Authorised Officers and Experts, 1996.
20) Exchange of Information between Member States of the EU on Routine Food Control Matters, 1996.

A3

Appendix 4

EEC food legislation

A Regulations

These are directly applicable to all member states and do not require national enactment. Certain measures may be adopted nationally to aid the enforcement or interpretation of the Regulations. Listed below are those Regulations which relate to the content of this book. Regulations are listed by the reference number first, followed by the year.

Subject	Regulation	Amendments
Butter – analysis for sitosterol and stigmasterol	86/94	
Butteroil – analysis for sitosterol and stigmasterol	3942/92	2539/93
Certificates of special character	2082/92	
Certificates of special character – rules	1848/93	2515/94
Contaminants – general	315/93	
Dairy designations – protection	1898/87	222/88
Eggs – marketing standards	1907/90	2617/93
Fats – spreadable	2991/94	
Geographical indications and designations of origin	2081/92	
Geographical indications and designations of origin – rules	2037/93	
Milk – analysis	2721/95	
Milk and milk products	1411/71	1556/74, 566/76, 222/88, 2138/92
MRLs in foods of animal origin	2377/90	762/92, 2701/94, 2703/94, 3059/94, 1102/95, 1441/95, 1442/95, 1798/95, 2796/95, 2804/95, 281/96, 282/96
Oils and fat – characteristics and analysis	2568/91	
Olive oil	2568/91	3682/91, 1429/92, 1683/92, 3288/92, 183/93, 177/94, 656/95, 2527/95
Olive oil – characteristics and analysis	2568/91	1683/92

(*continued*)

Subject	Regulation	Amendments
Organic production	2092/91	1535/92, 2083/92, 2608/93, 468/94, 1468/94, 2381/94, 529/95, 1201/95, 1202/95, 1935/95, 418/96
Organic production – date of application	3713/92	1593/93, 688/94, 2580/94
Organic production – imports – rules	94/92	
Organic production – imports – rules – inspection certificates	3457/92	
Organic production – rules	207/93	
Poultry – marketing standards	1906/90	
Poultry – marketing standards – rules	1538/91	2891/93
Radioactive contamination of foods	3954/87	2218/89
Radioactive contamination of foods – minor foods	944/89	
Sardines – preserved	2136/89	
Spirit drinks	1576/89	3280/92, 3378/94
Spirit drinks – additional rules	2675/94	
Spirit drinks – rules	1014/90	1180/91, 1781/91, 3458/92, 2675/94, 1712/95, 2626/95
Tuna and bonito – preserved	1536/92	
Water content of frozen poultry – standards	2967/76	1691/77, 641/79, 2632/86, 2835/80, 3204/83
Water content of frozen poultry – standards – rules	2785/80	3134/81

B Directives

These require enactment by national regulations for implementation. Given below are details of the major Directives relating to food law. Many of the Regulations summarised in this Guide implement the Directives listed. Directives are listed by the year first followed by the reference number.

Subject	Directive	Amendments
Additives – antioxidants – purity criteria	78/664	82/712
Additives – colours	94/36	
Additives – colours – purity criteria	95/45	
Additives – emulsifiers and stabilisers – purity criteria	78/663	82/504, 90/612, 92/4
Additives – general	89/107	94/34
Additives – miscellaneous	95/2	
Additives – purity criteria – analysis	81/712	
Additives – sweeteners	94/35	
Additives – sweeteners – purity criteria	95/31	
Aquaculture products – hygiene	91/67	93/54, 95/22, 95/70
Caseins and caseinates	83/417	

A4

Subject	Directive	Amendments
Caseins and caseinates – analysis	85/503	
Caseins and caseinates – sampling	86/424	
Cocoa and chocolate	73/241	74/411, 74/644, 75/155, 76/628, 78/609, 78/842, 80/608, 85/7, 89/344
Coffee	77/436	85/7
Coffee – analysis	79/1066	
Erucic acid	76/621	
Erucic acid – analysis	80/891	
Extraction solvents	88/344	92/115, 94/52
Fishery products – hygiene	91/493	95/71
Fishery products – hygiene – on board vessels	92/48	
Flavourings	88/388	
Foods for energy restricted diets	96/8	
Fruit juices	93/77	
Honey	74/409	
Hygiene of foodstuffs	93/43	
Hygiene of foodstuffs – derogation	96/3	
Infant formulae and follow-on formulae	91/321	96/4
Infant formulae and follow-on formulae – exports	92/52	
Jams, jellies, etc.	79/693	88/593
Labelling	79/112	93/102, 95/42
Labelling – additional indications	94/54	
Labelling – lot marking	89/396	91/238, 92/11
Live bivalve molluscs – hygiene	91/492	
Materials and articles in contact with food – ceramic articles	84/500	
Materials and articles in contact with food – general	89/109	
Materials and articles in contact with food – migration – simulants	85/572	
Materials and articles in contact with food – migration – testing	82/711	93/8
Materials and articles in contact with food – plastic	90/128	92/39, 93/9
Materials and articles in contact with food – plastics	90/128	92/39, 93/9, 95/3, 96/11
Materials and articles in contact with food – regenerated cellulose film	93/10	93/111
Materials and articles in contact with food – symbols	80/590	
Materials and articles in contact with food – vinyl chloride	78/142	
Materials and articles in contact with food – vinyl chloride – analysis – level of monomer in material	80/766	
Materials and articles in contact with food – vinyl chloride – analysis of food	81/432	

(continued)

Subject	Directive	Amendments
Meat – fresh	64/433	66/601, 73/101, 75/379, 83/90, 85/586, 85/3768, 87/3805, 89/662, 91/497, 92/5, 95/23
Meat – minced meat and meat preparations	94/65	
Meat products – amending	92/5	
Meat products – health	77/99	85/327, 85/586, 85/3768, 87/3805, 89/227, 89/662, 92/5, 92/45, 92/116, 92/118, 95/68
Milk – preserved and dried	76/118	78/630, 83/635
Milk – preserved and dried – analysis	79/1067	
Milk – preserved and dried – sampling	87/524	
Milk and milk products – health	92/46	92/118, 94/330, 94/71
Natural mineral waters	80/777	85/7
Nectars	93/45	
Official control	89/397	
Official control – additional measures	93/99	
Particular nutritional uses	89/398	
Plastic materials	90/128	95/3, 96/11
Poultry – health	71/118	73/101, 75/431, 78/50, 80/216, 84/335, 84/642, 85/326, 85/3768, 87/3805, 89/662, 92/116, 94/65
Poultry meat – amending	92/116	
Poultry meat – health	91/494	92/116, 93/121
Processed cereal based foods and baby foods	96/5	
Rabbit and farmed game meat – hygiene	91/495	92/65, 92/116, 94/65
Sampling and analysis – general	85/591	
Sugars	73/437	
Sugars – analysis	79/796	
Wild game meat – health	92/45	
Wild game meat – hygiene	92/45	92/116

Appendix 5

Useful addresses and internet sites

A   Addresses

Government departments

MAFF Consumer Helpline
Room 303a, Ergon House
17 Smith Square
London SW1P 3JR
Tel. 0345 573012 (all calls charged at local rate)
e-mail to: CONSUMER@INFO.MAFF.GOV.UK

Ministry of Agriculture, Fisheries and Food
Ergon House
c/o Nobel House
17 Smith Square
London
SW1P 3JR

Ministry of Agriculture, Fisheries and Food
(Meat Hygiene Division)
Tolworth Tower
Surbiton
Surrey
KT6 7DX

Department of Health
80 London Road
Elephant and Castle
London
SE1

Scottish Office
(Agriculture and Fisheries Department)
Pentland House
47 Robb's Loan
Edinburgh
EH14 1TW

A5

Department of Health and Social Services
(Food Control Branch)
Castle Buildings
Stormont
Belfast
BT4 3RA

Department of Agriculture for N.I.
(Food Policy Division)
Dundonald House
Upper Newtownards Road
Belfast
BT4 3SB

Trade Descriptions/Weights and Measures
Department of Trade and Industry
(Consumer Affairs Division)
1 Victoria Street
London
SW1H 0ET

EEC Commission (London office)
EEC Commission
8 Storey's Gate
London
SW1P 3AT

Enforcement
LACOTS
P.O. Box 6
Robert Street
Croydon
CR9 1LG

A5 *Her Majesty's Stationery Office/The Stationery Office*
Copies of the legislation listed in this guide may be obtained from:

Postal application:

The Stationery Office Books
P.O. Box 276
London
SW8 5DT

Personal application:

The Stationery Office
49 High Holborn
London
WC1V 6HB

Or through local branches of HMSO or through booksellers.

B Internet sites
The following should provide good access to relevant pages:

MAFF Home Page
http://www.maff.gov.uk

Department of Health Home Page
http://www.open.gov.uk/doh/dhhome.htm

Trading Standards Net
http://www.xodesign.co.uk/tsnet/

Food Law Pages, University of Reading
http://www.fst.rdg.ac.uk/people/ajukesdj/l-index.htm

European Commission – UK Office
http://www.cec.org.uk/

Institute of Food Science and Technology
http://www.easynet.co.uk/ifst/

Appendix 6

Late amendments

Late amendments

This Appendix provides some information concerning the regulations which were published between the date the original manuscript was completed (25th November 1996) and the date when the page proofs were checked (1st June 1997). The information is given in the order it would appear in the book.

A-Z food standards: Infant formula and follow-on formula

Regulation: Infant Formula and Follow-on Formula (Amendment) Regulations 1997 (1997/451)

These Regulations introduce additional nutritional labelling requirements including a new listing of reference values for certain vitamins and minerals. Modifications have been made to certain of the composition requirements listed in the Table on pages 42-44. These include:

changes to the protein requirements for infant formulae;
a change to the minimum lipid content of infant formulae;
additional requirements for lipids for both infant formulae and follow-on formulae;
control of the level of selenium in infant formulae;
substitution of controls for nicotinamide by niacin;
addition of controls on certain nucleotides.

Provision is also made for a claim relating to 'reduction of risk to allergy to milk proteins' and the necessary conditions to be met.
See the Regulations for specific details of these amendments.

A-Z food standards: Meat - Specified bovine material

Regulation: Specified Bovine Material Order 1997 (1997/617)

As anticipated on pages 51-53, the controls relating to specified bovine material have been updated and replaced. Those listed were originally

A6

replaced in late 1996 and have subsequently been further replaced by the above Order. Given the rapidly changing controls in this area, readers are advised to obtain the most recent copy of the Order for full details.

Additives: Sweeteners

Regulation: Sweeteners in Food (Amendment) Regulations 1997 (1997/814)

The amending Regulations provide additional controls relating to certain 'relevant compound foods' (also now defined). The main amendments however relate to the listing of foods in which sweeteners may be used and the level of use (as given on pages 125-135)

The food description 'Vitamins and dietary preparations' has been amended to 'Food supplements/diet integrators based on vitamins and/or mineral elements, syrup-type or chewable' and, in addition, the following sweeteners may now be used: E952, 1250 mg/kg; E954, 1200 mg/kg; E959, 400 mg/kg.

The following additional categories have been added:

Breakfast cereals with a fibre content of more than 15%, and containing at least 20% bran, energy reduced or with no added sugar: E950, 350 mg/kg; E951, 1000 mg/kg; E954, 100 mg/kg; E959, 50 mg/kg.

Breath-freshening micro-sweets, with no added sugar: E950, 2500 mg/kg; E951, 6000 mg/kg; E952, 2500 mg/kg; E954, 100 mg/kg; E959, 400 mg/kg.

Strongly flavoured freshening throat pastilles with no added sugar: E951, 2000 mg/kg.

Energy-reduced tablet form confectionery: E950, 500 mg/kg.

Cornets and wafers with no added sugar for ice-cream: E950, 2000 mg/kg; E954, 800 mg/kg; E959, 50 mg/kg.

Essoblaten: E950, 2000 mg/kg.

Drinks consisting of a mixture of a non-alcoholic drink and beer, cider, perry spirits or wine: E950, 350 mg/l; E951, 600 mg/l; E952, 250 mg/l; E954, 80 mg/l; E959, 30 mg/l.

Energy-reduced beer: E950, 25 mg/l, E951, 25 mg/l; E959, 10 mg/kg.

Spirit drinks containing less than 15% alcohol by volume: E950, 350 mg/kg, E951, 600 mg/kg; E954, 80 mg/kg; E959, 30 mg/kg.

'Feinkostsalat': E950, 350 mg/kg; E951, 350 mg/kg; E954, 160 mg/kg; E959, 50 mg/kg.

Energy-reduced soups: E950, 110 mg/l; E951, 110 mg/l; E954, 110 mg/l; E959, 50 mg/l.

Edible ices, energy-reduced or with no added sugar: E957, 50 mg/kg.

"Snacks": certain flavours of ready to eat, prepacked, dry, savoury starch products and coated nuts: E959, 50 mg/kg.

Complete formulae and nutritional supplements for use under medical supervision: E959, 100 mg/kg.

Contaminants: Pesticides

Regulation: Pesticides (Maximum Residue Levels in Crops, Food and Feeding
   Stuffs) (Amendment) Regulations 1997 (1997/567)

The amendments add controls on the following pesticides (to those already
listed on page 144):
   Chlormequat, Disulfoton, Fenbutatin oxide, Fentin, Mecarbam, Phorate,
   Propoxur, Propyzamide, Triforine.
Certain pesticides have their uses restricted or altered to take into account
Community measures now agreed.

Processing and packaging: Plastic materials and articles in contact with food

Regulation: Plastic Materials and Articles in Contact with Food (Amendment)
   (No.2) Regulations 1996 (1996/2817)

The Regulations update the list of permitted monomers in plastic materials
and articles for use in contact with food. Reference is made to permitted
monomers on page 159. However, due to the complexity of the lists, no
details are given and the Regulations should be consulted for details.

Labelling: Organic foods

Regulation: Organic Products (Amendment) Regulations 1997 (1997/166)

The amendment provides that the operator shall include on the labelling a
reference to the code number of the relevant inspection authority or body.

Index

Note: Entries in bold have their own sub-section in the book

253